BestMasters

Mit „**BestMasters**" zeichnet Springer die besten Masterarbeiten aus, die an renommierten Hochschulen in Deutschland, Österreich und der Schweiz entstanden sind. Die mit Höchstnote ausgezeichneten Arbeiten wurden durch Gutachter zur Veröffentlichung empfohlen und behandeln aktuelle Themen aus unterschiedlichen Fachgebieten der Naturwissenschaften, Psychologie, Technik und Wirtschaftswissenschaften. Die Reihe wendet sich an Praktiker und Wissenschaftler gleichermaßen und soll insbesondere auch Nachwuchswissenschaftlern Orientierung geben.

Springer awards "**BestMasters**" to the best master's theses which have been completed at renowned Universities in Germany, Austria, and Switzerland. The studies received highest marks and were recommended for publication by supervisors. They address current issues from various fields of research in natural sciences, psychology, technology, and economics. The series addresses practitioners as well as scientists and, in particular, offers guidance for early stage researchers.

Sophia Denker

Characterizing Multiparticle Entanglement Using the Schmidt Decomposition of Operators

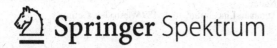

Springer Spektrum

Sophia Denker
Hövels, Germany

ISSN 2625-3577   ISSN 2625-3615   (electronic)
BestMasters
ISBN 978-3-658-43202-7   ISBN 978-3-658-43203-4   (eBook)
https://doi.org/10.1007/978-3-658-43203-4

This Springer Spektrum imprint is published by the registered company Springer Fachmedien
Wiesbaden GmbH, part of Springer Nature.
The registered company address is: Abraham-Lincoln-Str. 46, 65189 Wiesbaden, Germany

Paper in this product is recyclable.

Abstract

Witness operators are a useful tool to detect and quantify entanglement. A standard way to construct them is based on the fidelity of pure states and mathematically relies on the Schmidt decomposition of vectors [31]. In this thesis a method to build entanglement witnesses using the Schmidt decomposition of operators is presented. One can show that these are strictly stronger than the fidelity witnesses. Moreover, the concept can be generalized easily to the multipartite case and one may use it to quantify the dimensionality of entanglement. Finally, this scheme will be used to provide two algorithms that can be combined in order to improve given witnesses for multiparticle entanglement.

Contents

Introduction

In the 20th century it turned out that classical mechanics was not sufficient anymore in order to describe certain phenomena in physics. Hence, in the 1920s the theory of quantum mechanics was developed, which provides a mathematical framework to describe for example physical interactions on small energy scales. In 1935 the phenomenon of entanglement was first described by Einstein, Podolsky and Rosen (EPR) [49] and Schrödinger [52], who questioned the completeness of the theory. The assumption arose that there are hidden variables determining the physical happening. In 1964 Bell found that if there were such hidden variables, the correlations between the results of measurements of composed systems would be upper-bounded by a certain number. However, using entanglement, quantum mechanics violates this bound and thus, the assumption that there are hidden variables completing the theory of quantum mechanics may not be true. This violation could first be shown experimentally by John F. Clauser et al. in 1972 [13]. Later, Alain Aspect et al. performed several experiments closing one of two loopholes in Clauser's experiment [15–17]. Finally, in 2015 the first loophole free experiments were done simultaneously by two groups lead by Anton Zeilinger [27, 56, 61]. For their work, these three physicists received the Nobel Prize in Physics 2022.

With the development of quantum mechanics, also quantum information theory arose, where entanglement is a topic of great interest, too, as it is a resource for many applications such as quantum teleportation, quantum cryptography and quantum metrology. Consequently, entanglement detection is as important. One useful tool to detect and quantify it are entanglement witnesses. These are observables and hence can be implemented experimentally, which motivates their investigation. In this thesis a new type of witness, which is based on the Schmidt decomposition in the operator space (OSD), is introduced. One can show that these witnesses are

© The Author(s), under exclusive license to Springer Fachmedien Wiesbaden GmbH, part of Springer Nature 2023
S. Denker, *Characterizing Multiparticle Entanglement Using the Schmidt Decomposition of Operators*, BestMasters,
https://doi.org/10.1007/978-3-658-43203-4_1

strictly stronger than those, based on the Schmidt decomposition in the vector space, which are considered as the standard witnesses. Furthermore the concept can be generalized to the multiparticle case and also be used to quantify the dimensionality of entanglement.

In the first part of this thesis the physical and mathematical background is given. We start with putting the concept of states and measurements into mathematical terms. After that, quantum entanglement will be discussed in detail and having defined it for the bipartite case, some examples for detection criteria are given. Moreover the Schmidt rank and Schmidt number as quantity for the dimensionality of entanglement are introduced and further, quantum entanglement in the multipartite case will be discussed. Next, entanglement witnesses are introduced. Their formal definition will be given as well as some examples how to construct and quantify them. Additionally, Schmidt number witnesses and witnesses for multiparticle entanglement will be explained. Lastly, some examples for applications of quantum entanglement are given, which are quantum teleportation, quantum cryptography and quantum metrology.

Having introduced the preliminaries, in the next chapter the new type of witnesses, based on the Schmidt decomposition of operators is discussed. First, we will explain how to construct them and further, we will show that they are indeed strictly stronger than the standard witnesses, which are based on the vector Schmidt decomposition. After that, two algorithms to improve given entanglement witnesses or find an entanglement witness that detects a certain target state are constructed. One is based on optimizing the operator Schmidt coefficients (OSC) and the other on optimizing the Schmidt operators (SO). Further, we will give an example where those algorithms are applied to a PPT entangled state. It is found that, starting with a completely random input, one can only find the best witness if both algorithms are applied. Moreover, we will show that the algorithms behave as expected and therefore can be generalized to multiparticle systems.

The next chapter deals with the generalization of the OSD-based witness to the multipartite case. First, its construction will be addressed. Then, we will adapt the two optimization algorithms to the multiparticle case. Starting with the optimization with respect to the operator Schmidt coefficients, we will show that the algorithm improves the fidelity witness quite well for many states. However, we will find that for W states the optimization does not work that well. Therefore, two ideas to modify the algorithm in order to find better results are introduced and it is found that indeed those modifications lead to better results for the W states. Furthermore, another optimization approach will be discussed. The second part of this chapter focuses on the optimization algorithm with respect to the Schmidt operators. After the adaption, it will be applied to the same example states as from the previous part. However, one finds that this optimization algorithm is not as effective as the previous one. Therefore, we will discuss some strategies to combine it with the operator Schmidt coefficient optimization. Lastly, having found the best strategy, the witnesses will be improved further and compared to other results.

In Chapter 5 we will show how to construct Schmidt number witnesses and an example will demonstrate that they can detect higher entanglement dimensionalities than the fidelity-based Schmidt number witnesses.

Finally, in Chapter 6 a summary of all results will be given and some open questions will be discussed.

Physical and Mathematical Background

2

2.1 States and Measurements

Every physical experiment can be interpreted as a preparation and a measurement, where a state is prepared, an observable measured and an outcome obtained. In the next two subsections these expressions are put into mathematical terms.

2.1.1 States

According to one of the postulates of quantum mechanics, the state of a system can be described by the elements of a *Hilbert space* [42]. A Hilbert space $\mathcal{H}$ is a complete d-dimensional complex vector space whose norm is induced by a scalar product. Therefore, the vector $|\psi\rangle \in \mathcal{H}$ with $\sqrt{\langle\psi|\psi\rangle} = 1$ is a *quantum state*. If the state of a system can be described by a single vector $|\psi\rangle$, it is called *pure*.

However, considering the situation where a source produces several pure states $|\psi_i\rangle$ with certain probabilities p_i, the description needs to be generalized. In order to do so, the *density matrix* is defined:

$$\rho = \sum_i p_i |\psi_i\rangle\langle\psi_i|. \tag{2.1}$$

The probability for the system's actual state is now encoded in the density matrix. Since the values $\{p_i\}$ form a probability distribution and $(|\psi_i\rangle\langle\psi_i|)^\dagger = |\psi_i\rangle\langle\psi_i|$, every density matrix has the following properties:

© The Author(s), under exclusive license to Springer Fachmedien Wiesbaden GmbH, part of Springer Nature 2023
S. Denker, *Characterizing Multiparticle Entanglement Using the Schmidt Decomposition of Operators*, BestMasters,
https://doi.org/10.1007/978-3-658-43203-4_2

$$\rho = \rho^\dagger \tag{2.2}$$

$$\text{Tr}(\rho) = 1 \tag{2.3}$$

$$\rho \geq 0. \tag{2.4}$$

That is, it is hermitian, normalized and has only non-negative eigenvalues. For a pure state, the density matrix is simply the projector $|\psi\rangle\langle\psi|$ of its state vector $|\psi\rangle$.

The set of all density matrices is denoted *state space*. Due to the definition of the density matrix, the convex combination of two states $\rho = p\rho_1 + (1-p)\rho_2$ with $p \in [0,1]$ is a state again. Hence, the state space is a *convex set*. Pure states are the extreme points and therefore on the boundary of the set, all other states are called *mixed states*. Examples of convex sets are shown in Figure 2.1.

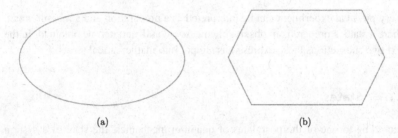

(a)   (b)

Figure 2.1 Two examples for convex sets. The extreme points of set (a) are all points on the boundary, set (b) has its extreme points at the corners

2.1.2   Measurements

Having described the states of a system, we now want to formulate *measurements* mathematically, which are also defined by the axioms of quantum mechanics. There is to mention that many textbooks only introduce *projective* or *von-Neumann measurements*, which are special cases of quantum measurements [42]. In the following, those projective measurements, as well as *positive operator valued measures* as a generalization are explained.

Observables and Projective Measurements

We start by considering pure states again. One axiom of quantum mechanics states that measurements are described by hermitian bounded operators, called observables [5]:

$$M = M^\dagger = \sum_k \lambda_k |e_k\rangle\langle e_k|, \tag{2.5}$$

where the λ_k are the eigenvalues and $|e_k\rangle$ are the eigenvectors of M. Since M is hermitian, its eigenvalues λ_k are real; they denote the possible outcomes of the measurement. The eigenvectors are orthonormal, i. e. they fulfill $\langle e_k|e_l\rangle = \delta_{kl}$ and therefore the probability to obtain outcome λ_i is given by $p_i = |\langle\psi|e_i\rangle|^2$. After a measurement, the state $|\psi\rangle$ is projected to the eigenstate $|e_i\rangle$ corresponding to the obtained outcome λ_i. The mean value is then

$$\langle M\rangle = \langle\psi|M|\psi\rangle \tag{2.6}$$

Extending this to the case of the density matrix, the expectation value can be computed using the trace:

$$\langle M\rangle_\rho = \sum_k p_k \langle\psi_k|M|\psi_k\rangle \tag{2.7}$$

$$\langle M\rangle_\rho = \mathrm{Tr}\left(M\sum_k p_k|\psi_k\rangle\langle\psi_k|\right) \tag{2.8}$$

$$\langle M\rangle_\rho = \mathrm{Tr}(M\rho). \tag{2.9}$$

These measurements are called *projective measurements* or *von-Neumann measurements* and have the property that every outcome can be reproduced if one measurement is performed directly after the other. A more general description of measurements is given by positive operator valued measures, which are introduced in the following.

Positive Operator Valued Measures

As mentioned before, the concept of projective measurements can be generalized to *positive operator valued measures (POVMs)*. A POVM is a set of *effects* $\{E_i\}$. These are given by $E_i = M_i M_i^\dagger = M_i^\dagger M_i$, where the $\{M_i\}$ are *measurement operators*. Since we want to describe probabilities, the effects need to be positive and sum up to one:

$$E_i \geq 0 \tag{2.10}$$

$$\sum_i E_i = \mathbb{1}. \tag{2.11}$$

With this we can describe measurements, where the probability to obtain outcome i is given by $p_i = \mathrm{Tr}(E_i \rho)$ and the state after the measurement is proportional to $\rho' = M_i \rho M_i^\dagger$. If all effects are projectors, we have a *projective valued measure (PVM)*, which is therefore simply a special case of a POVM.

In practice, a PVM is more easy to implement. Thus, it is desirable to describe measurements as PVMs. For this, the *Naimark dilation* is useful, which states that every POVM is a PVM on a higher dimensional system. One can find the proof in [44] for example.

2.2   Quantum Entanglement

When we have two or more particles, the considerations in Section 2.1 are not sufficient anymore and therefore, we need a way to describe *composed systems*. In order to do so, we use the *tensor product*: If each quantum state $|\psi_n\rangle$ lives in its own Hilbert space $\mathcal{H}_n$, the composed system can then be described as

$$|\Psi_{12\ldots N}\rangle = |\psi_1\rangle \otimes |\psi_2\rangle \otimes \ldots \otimes |\psi_N\rangle \in \mathcal{H}_1 \otimes \mathcal{H}_2 \otimes \ldots \otimes \mathcal{H}_N \qquad (2.12)$$

and $|\Psi_{12\ldots N}\rangle$ is called *product state*. However, in quantum mechanics it is not always the case that a state $|\Psi\rangle$ living on a composed system can be written as in Equation (2.12), which describes the phenomenon of entanglement [24]. In the following, we will take a closer look at it, starting with the case of two particles and afterwards considering multiparticle systems.

2.2.1   Bipartite Entanglement

Mathematically, bipartite entanglement for pure states is defined as follows [19]:

Definition 2.1 (Entanglement for pure states). *Having a bipartite system, shared by the two parties Alice (A) and Bob (B), every pure state $|\Psi_{AB}\rangle \in \mathcal{H}_{AB}$ that can be written as*

$$|\Psi_{AB}\rangle = |\psi_A\rangle \otimes |\psi_B\rangle \qquad (2.13)$$

with $|\psi_A\rangle \in \mathcal{H}_A$ and $|\psi_B\rangle \in \mathcal{H}_B$ is called separable. States that can not be written in this form are called entangled.

As well as in Section 2.1, this definition can be extended for mixed states using the density matrix. We obtain:

Definition 2.2 (Entanglement for mixed states). *A bipartite mixed state ρ is separable if it can be written as*

$$\rho_{\text{sep}}^{AB} = \sum_k p_k \rho_k^A \otimes \rho_k^B, \qquad (2.14)$$

where all p_k are positive and sum up to one. Otherwise it is entangled.

Physically this means that separable states can be generated by *local operations and classical communication (LOCC)*. That is that Alice and Bob prepare states $\rho_k^A \otimes \rho_k^B$ by local operations and then communicate classically and agree on probabilities p_k resulting in the state given by Equation (2.14). Hence, entangled states are those, which can not be created by LOCC. Furthermore, analogous to the definition of pure states, $\rho = \rho^A \otimes \rho^B$ is called *product state*.

In the state space for bipartite states we differ now between *entangled* and *separable* states. From their definition we find that separable states form a convex subset of the entire state space. Furthermore, the pure product vectors are the extreme points of not only the separable states but also of the whole state space. Figure 2.2 shows a schematic sketch of the state space [19].

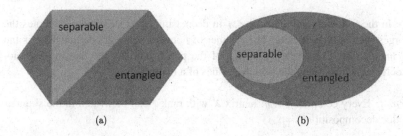

(a)   (b)

Figure 2.2 Schematic sketch of the state space for bipartite states. Figure (a) shows the separable states as a convex subset of all states where the product states are extreme points of the total set. Figure (b) is a more common representation of the bipartite state space

Having defined entanglement, we now want to learn more about its detection and quantification. Hence, the Schmidt decomposition, the PPT criterion and the

CCNR criterion will be discussed in the following. However, it is to mention that there are many more criteria to detect entanglement [19].

The Schmidt Decomposition

The Schmidt decomposition is useful to detect and quantify entanglement of pure states. It is given by [19]:

Lemma 2.1 (Schmidt decomposition for vectors). *Let* $|\Psi\rangle = \sum_{i,j=1}^{d_A,d_B} c_{ij}|a_i\rangle \otimes |b_j\rangle \in$ $\mathcal{H}_A^{d_A} \otimes \mathcal{H}_B^{d_B}$ *be a state vector in a bipartite system, then there exist orthonormal bases (ONB)* $\{|\alpha_i\rangle\}$ *and* $\{|\beta_j\rangle\}$ *of* $\mathcal{H}_A^{d_A}$ *and* $\mathcal{H}_B^{d_B}$, *respectively and some real coefficients* $s_k \geq 0$ *such that* $|\Psi\rangle$ *can be written as*

$$|\Psi\rangle = \sum_{k=1}^{d} s_k|\alpha_k\beta_k\rangle, \tag{2.15}$$

where $d = \min(d_A, d_B)$ *denotes the minimum dimension of the two Hilbert spaces and* $|\alpha_k\beta_k\rangle = |\alpha_k\rangle \otimes |\beta_k\rangle$ *is a short hand notation for the tensor product. The* s_k *are unique and called Schmidt coefficients. Furthermore, they fulfill* $\sum_{k=1}^{d} s_k^2 = 1$ *and if the Schmidt coefficients differ pairwise, the basis vectors* $|\alpha_k\rangle$ *and* $|\beta_k\rangle$ *are unique as well.*

In the following we define the s_k in decreasing order such that s_1 denotes the largest Schmidt coefficient. The number of nonzero Schmidt coefficients r is the *Schmidt rank*. Considering a sketch of the proof, one can see that the Schmidt coefficients are simply the singular values of a matrix (c_{ij}).

Proof. Every complex $n \times m$ matrix X with rank r can be written in the singular value decomposition:

$$X = USV^{\dagger}, \tag{2.16}$$

where U is a unitary $n \times n$ and V a unitary $m \times m$ matrix. The matrix S is diagonal and of the form $n \times m$. The coefficients on its diagonal are the *singular values* and r of them are nonzero.

Knowing this, we consider again the state $|\Psi\rangle = \sum_{i,j=1}^{d_A,d_B} c_{ij}|a_i\rangle \otimes |b_j\rangle$, writing the matrix (c_{ij}) in the singular value decomposition. Then, we obtain:

$$|\Psi\rangle = \sum_{ij} c_{ij}|a_i\rangle \otimes |b_j\rangle \tag{2.17}$$

$$= \sum_{ijk} u_{ik}s_k v_{kj}^\dagger |a_i\rangle \otimes |b_j\rangle \tag{2.18}$$

$$= \sum_{ijk} s_k u_{ik}|a_i\rangle \otimes v_{kj}^\dagger |b_j\rangle \tag{2.19}$$

$$= \sum_k s_k \left(\sum_i u_{ki}^T |a_i\rangle\right) \otimes \left(\sum_j v_{kj}^\dagger |b_j\rangle\right) \tag{2.20}$$

$$= \sum_k s_k |\alpha_k\rangle \otimes |\beta_k\rangle, \tag{2.21}$$

which is the Schmidt decomposition with r nonzero Schmidt coefficients s_k. $\square$

Considering Equation (2.15), the dimensionality of entanglement can be characterized. As mentioned before, the number of nonzero Schmidt coefficients r denotes the Schmidt rank, which means that if the Schmidt rank is $r = 1$, the state is separable. An easier way to determine the Schmidt coefficients is to use the fact that the singular values of a matrix C are the square root of the eigenvalues of a matrix $CC^\dagger$.

As before, the concept of the Schmidt rank can also be generalized to mixed states or rather density matrices [57].

Definition 2.3 (Schmidt number). *Consider the density matrix ρ of a mixed state $\rho = \sum_i p_i |\psi_i^{r_i}\rangle\langle\psi_i^{r_i}|$, where r_i is the rank of the state $|\psi_i^{r_i}\rangle$, in all possible decompositions. Then the Schmidt number k of the density matrix ρ is given by*

$$k = \min_{\text{all decompositions}} \left(\max_i (r_i)\right). \tag{2.22}$$

Denoting the set of density matrices with Schmidt number k or less by S_k, it is found that S_k is a convex compact subset of the set of all density matrices and fulfills $S_{k-1} \subset S_k$ [57]. Moreover, as in the case of pure states, having Schmidt number $k = 1$ means that the state is separable (see Figure 2.3).

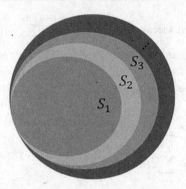

Figure 2.3 Schematic picture of the bipartite state space showing the subsets S_k of k-dimensionally entangled states. Note that S_1 is the set of separable states

The Peres-Horodecki Criterion

The *Peres-Horodecki criterion*, also known as *positive partial transpose (PPT) criterion* is a powerful tool to determine entanglement of bipartite states. In order to write down the PPT criterion, we first need to define the *partial transpose*.

Definition 2.4 (Partial transpose). *Considering a bipartite state ρ it can be written in some operator basis $\rho = \sum_{ijkl} \rho_{ijkl} |i\rangle\langle j| \otimes |k\rangle\langle l|$. The partial transpose is then defined as*

$$\rho^{T_A} = \sum_{ijkl} \rho_{ijkl} (|i\rangle\langle j|)^T \otimes |k\rangle\langle l| = \sum_{ijkl} \rho_{ijkl} |j\rangle\langle i| \otimes |k\rangle\langle l| \qquad (2.23)$$

$$\rho^{T_B} = \sum_{ijkl} \rho_{ijkl} |i\rangle\langle j| \otimes (|k\rangle\langle l|)^T = \sum_{ijkl} \rho_{ijkl} |i\rangle\langle j| \otimes |l\rangle\langle k|. \qquad (2.24)$$

Consequently, a state is *PPT* means that ρ^{T_A} and ρ^{T_B} both have only non-negative eigenvalues. With this, the PPT criterion can be formulated [22, 45].

Theorem 2.1 (PPT criterion). *If a bipartite state ρ is separable, it is PPT.*

Proof. Consider the partial transpose of a separable state:

$$\rho^{T_A} = \sum_k p_k \rho_k^T \otimes \rho_k \qquad (2.25)$$

$$= \sum_k p_k \rho_k' \otimes \rho_k \geq 0. \qquad (2.26)$$

Where we used that a transposed state is a state as well. The same can be shown analogously for ρ^{T_B}. □

Therefore, finding one negative eigenvalue of a partially transposed state ρ is sufficient to decide entanglement. These states are called *NPT* (negative partial transpose).

The other direction, in general, is not true, since there are also entangled states with positive partial transpose (see Figure 2.4). Those states are called *PPT entangled states*. However, Michał, Paweł and Ryszard Horodecki showed that for 2×2 and 2×3 systems the PPT criterion is sufficient for separability as well [22].

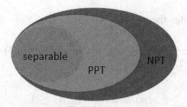

Figure 2.4 Sketch of the set of all states, showing that entangled states can be PPT or NPT

The Computable Cross Norm or Realignment (CCNR) Criterion

Another criterion to detect entanglement is the *CCNR criterion*. It has two names, since is was formulated independently in two different ways [19]. One way makes use of the cross norm of density matrices [50] and the other one was found realigning the entries of density matrices and computing their trace norm [8]. In order to understand the CCNR criterion and also being able to formulate it in a way such that it can be computed easily, first, the *Schmidt decomposition of operators (OSD)* is introduced [41].

Lemma 2.2 (Schmidt decomposition for operators). *Every operator X can be decomposed in the following way:*

$$X = \sum_{k=1}^{d^2} \mu_k G_k^A \otimes G_k^B, \tag{2.27}$$

where the $\mu_k \geq 0$ are called operator Schmidt coefficients and the $G_k^A \in \mathcal{H}_A^{d_A}$ and $G_k^B \in \mathcal{H}_B^{d_B}$ are hermitian and form orthonormal bases of their respective Hilbert *space fulfilling $Tr(G_i G_j) = \delta_{ij}$. The dimension d is given by $d = \min(d_A, d_B)$. As in the Schmidt decomposition of vectors (Lemma 2.1), the μ_k are unique and if they differ pairwise, the basis operators G_k^A and G_k^B are unique as well.*

Analogue to the vector Schmidt coefficients, we define the operator Schmidt (OS) coefficients in decreasing order. Furthermore, they are again singular values of a matrix M (or square rooted eigenvalues of $M M^\dagger$), where $M = (m_{ij})$ is found by decomposing the operator X in an arbitrary basis

$$X = \sum_{ij} m_{ij} H_i^A \otimes H_j^B. \tag{2.28}$$

The CCNR criterion states then:

Theorem 2.2 (CCNR criterion). *A state ρ is separable only if the sum of its operator Schmidt coefficients μ_k is smaller than one:*

$$\rho \text{ separable } \Rightarrow \sum_k \mu_k \leq 1. \tag{2.29}$$

The proof makes use of the fact that the sum over the Schmidt coefficients is a norm on the set of density matrices [19]. Hence, in the following proof we denote $\|\rho\| = \sum_k \mu_k$.

Proof. A pure product state $\rho = \rho_A \otimes \rho_B$ has only one nonzero Schmidt coefficient $\mu_1 = 1$, therefore $\|\rho\| = 1$ and the statement is proven for the trivial case. Consider now the norm of a general product state:

$$\|\rho\| = \left\| \sum_k p_k \rho_k^A \otimes \rho_k^B \right\| \tag{2.30}$$

$$\leq \sum_k p_k \|\rho_k^A \otimes \rho_k^B\| \tag{2.31}$$

$$= \sum_k p_k \tag{2.32}$$

$$= 1, \tag{2.33}$$

where in the first step the triangle inequality was used and in the second step that a general product state is a convex combination of pure product states. ☐

Therefore, if the sum of all operator Schmidt coefficients of a state ρ is larger than one, it is entangled.

Reformulation of PPT and CCNR Criterion

Lastly, there is to mention that the CCNR criterion complements the PPT criterion. This can be seen by reformulating the two criteria using the fact that any state can be decomposed in the following from:

$$\rho = \sum_{ijkl} \rho_{ijkl} |i\rangle\langle j| \otimes |k\rangle\langle l| \tag{2.34}$$

Then ρ_{ijkl} can be interpreted as a $d_A^2 \times d_B^2$ matrix and for pure states it fulfills that its trace norm is one:

$$\|\rho_{ijkl}\|_1 = \text{Tr}(\sqrt{\rho_{ijkl}\rho_{ijkl}^\dagger}) = 1 \tag{2.35}$$

Since any permutation $\pi(ijkl)$ of the indices gives a matrix that also has trace norm one, the separability criteria can be formulated in the following way [21]:

$$\rho \text{ separable} \Rightarrow \|\rho_{\pi(ijkl)}\|_1 \leq 1 \tag{2.36}$$

It turns out that there are only two inequivalent permutations that give separability criteria. One is obtained by permuting k and l which corresponds to the partial transposition of ρ and therefore to the PPT criterion:

$$\rho \text{ separable} \Rightarrow \|\rho_{ij\overset{\frown}{lk}}\|_1 \leq 1. \tag{2.37}$$

And the other one corresponds to the CCNR criterion:

$$\rho \text{ separable} \Rightarrow \|\rho_{i\overset{\frown}{kjl}}\|_1 \leq 1. \tag{2.38}$$

However, there are still states that are not detected by none of these criteria [51].

2.2.2   Multipartite Entanglement

In order to introduce multipartite entanglement, we start by considering three parties first. Then, in the second part it is shown that these considerations can be generalized for N parties.

Entanglement for Three Parties

Considering three parties Alice (A), Bob (B) and Charlie (C) sharing a pure state, six situations can occur. First, they could share a *fully separable* state, which means that there is no entanglement at all. Second, there could be the situation that there is only entanglement between two parties, for example Alice and Bob, whereas Charlies state is separated from the others. This is a *biseparable* state with respect to *bipartition AB|C*. Lastly, there could be entanglement between all three parties, which is called *genuine tripartite entanglement* and in the case of three qubits[1] can occur in two inequivalent ways [59].

Mathematically, those situations are expressed by equivalence classes, where an equivalence class contains all states that are equivalent. Thus, we need to know what it means that two states are *equivalent*. Some states can be obtained from others by LOCC, which means that each party can apply operators to their local system whereby they can communicate classically with each other. When the achievement to obtain one state from the other is not demanded with certainty, we have *stochastic LOCC (SLOCC)*. This can be used to define the equivalence of two states [59].

Definition 2.5 (Equivalence of states). *Two states $|\Psi\rangle$ and $|\Phi\rangle$ are equivalent if $|\Psi\rangle$ can be converted into $|\Phi\rangle$ via SLOCC and vice versa. For three-qubit states this means they are equivalent if and only if invertible operators A, B and C exist, such that*

$$|\Phi\rangle = A \otimes B \otimes C |\Psi\rangle \qquad (2.39)$$

$$|\Psi\rangle = A^{-1} \otimes B^{-1} \otimes C^{-1} |\Phi\rangle. \qquad (2.40)$$

Knowing this, the six equivalence classes and one representative example of these can be given, according to [59]:

---

[1] Qubits are states living in a two-dimensional Hilbert space. In the following, if nothing else is noted, states are referred to as qubits.

product states: $|\Psi_{ABC}\rangle = |0\rangle_A \otimes |0\rangle_B \otimes |0\rangle_C$ (2.41)

biseparable $A|BC$: $|\Psi_{A|BC}\rangle = \frac{1}{\sqrt{2}}|0\rangle_A \otimes (|00\rangle_{BC} + |11\rangle_{BC})$ (2.42)

biseparable $C|AB$: $|\Psi_{C|AB}\rangle = \frac{1}{\sqrt{2}}|0\rangle_C \otimes (|00\rangle_{AB} + |11\rangle_{AB})$ (2.43)

biseparable $B|AC$: $|\Psi_{B|AC}\rangle = \frac{1}{\sqrt{2}}|0\rangle_B \otimes (|00\rangle_{AC} + |11\rangle_{AC})$ (2.44)

entangled (W class): $|W\rangle = \frac{1}{\sqrt{3}}(|001\rangle_{ABC} + |010\rangle_{ABC} + |100\rangle_{ABC})$

                                                                              (2.45)

entangled (GHZ class): $|GHZ\rangle = \frac{1}{\sqrt{2}}(|000\rangle_{ABC} + |111\rangle_{ABC})$ (2.46)

It was found that every pure three-qubit state vector, up to local unitaries, can be written as

$$|\Psi_{GHZ}\rangle = \lambda_0|000\rangle + \lambda_1 e^{i\theta}|100\rangle + \lambda_2|101\rangle + \lambda_3|110\rangle + \lambda_4|111\rangle, \quad (2.47)$$

with $\lambda_i \geq 0$ fulfilling $\sum_i \lambda_i^2 = 1$ and $\theta \in [0, \pi]$. These states are called Greenberger-Horne-Zeilinger (GHZ) type states and if $\theta \neq 0$ or $\lambda_4 \neq 0$, they describe one from of genuine tripartite entanglement. However, as mentioned before, three qubits can be entangled in two different ways. The second type of entangled states are W type vectors, which are expressed as

$$|\Psi_W\rangle = \lambda_0|000\rangle + \lambda_1|100\rangle + \lambda_2|101\rangle + \lambda_3|110\rangle. \quad (2.48)$$

One can see that the W type vectors are of the same form as GHZ type vectors but with $\theta = \lambda_4 = 0$. So, by adding an infinitesimal λ_4 term to a W type state, one can always find a GHZ type state close to it. Thus, the W type vectors form a subset of the GHZ type vectors. The two representatives of the W class and the GHZ class given in equations (2.45) and (2.46) are called W state and GHZ state [59].

In [30], Acín, Bruß, Lewenstein and Sanpera extended the previously described classification to mixed states. The classes are defined in the following way:

i) The class S contains all separable states, i. e. states, that can be written as convex combination of projectors onto product vectors: $\rho_{\text{sep}} = \sum_i p_i \rho_i^A \otimes \rho_i^B \otimes \rho_i^C$.

ii) The class B contains all biseparable states, i. e. states, that can be written as convex combination of projectors onto product vectors and biseparable vectors with respect to all bipartitions:
$$\rho_{bs} = \sum_i p_i \rho_i^A \otimes \rho_i^{BC} + \sum_i q_i \rho_i^B \otimes \rho_i^{AC} + \sum_i r_i \rho_i^C \otimes \rho_i^{AB}.$$

iii) The W class contains all states, that can be written as convex combination of projectors on product, biseparable and W type vectors.

iv) Lastly, the GHZ class contains all physical states.

These sets are convex, compact and embedded into each other, which means $S \subset B \subset W \subset GHZ$. In Figure 2.5, a schematic picture of the sets is shown.

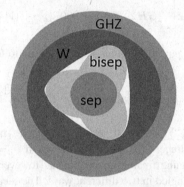

Figure 2.5 Schematic sketch of the set of all three-qubit states. The class S is here denoted by *sep* and B by *bisep*. Note that there exist states, which are biseparable with respect to all three bipartitions but not fully separable

Entanglement for N Parties

As well as for three parties, entanglement classes can also be defined for multiple parties. According to [19] there are several possibilities to classify N-qubit states but here we want to restrict to the following:

Considering a pure N-partite state $|\Psi\rangle$, it is fully separable if it can be written as

$$|\Psi\rangle = |\psi_1\rangle \otimes |\psi_2\rangle \otimes ... \otimes |\psi_N\rangle. \tag{2.49}$$

If it can not be written in this form, it contains some entanglement [19]. It is *m-separable* if the N parties can be split into $1 < m < N$ partitions P_i with

$$|\Psi\rangle = |\Psi_{P_1}\rangle \otimes |\Psi_{P_2}\rangle \otimes \ldots \otimes |\Psi_{P_m}\rangle. \tag{2.50}$$

Analogue to the previous sections, these classes can also be generalized to mixed states by taking convex combinations of pure projective states. Where for m-separable states, there is to mention that the states in the convex combination might belong to different partitions. States which are neither fully separable nor m-separable are called *truly N-partite entangled* [19] or *genuine multipartite entangled (GME)*. However, different from the tripartite case, it is not easy to define equivalence classes via SLOCC for GME. This is due to the fact that already for four qubits there are infinitely many SLOCC equivalence classes [19].

2.3 Entanglement Witnesses

A useful tool to detect and quantify entanglement in experiments are *entanglement witnesses*. These observables can be defined for the bipartite, as well as for the multipartite case. Moreover, to characterize the *dimensionality* of entanglement, *Schmidt witnesses* can be introduced.

2.3.1 Bipartite Entanglement Witnesses

For bipartite states an entanglement witness is an observable defined as follows [19]:

Definition 2.6 (Entanglement witness). *A hermitian operator $\mathcal{W}$ is called entanglement witness if*

$$Tr(\mathcal{W}\sigma) \geq 0 \; \forall \sigma \; separable \tag{2.51}$$

and it exists at least one entangled state ρ such that

$$Tr(\mathcal{W}\rho) < 0. \tag{2.52}$$

This means that if the expectation value for some state ρ_{det} is negative, one can conclude that it is entangled. Thus, the state ρ_{det} is *detected* by the witness.

Geometrically, entanglement witnesses can be interpreted as a hyperplane in the set of states. That is, all states $\tilde{\rho}$ fulfilling $Tr(\mathcal{W}\tilde{\rho}) = 0$ form a hyperplane separating the set of bipartite states in two parts. One part contains all states with negative expectation values, which are, per construction, entangled states and the other part contains all states with positive expectation value, including all separable states. According to the Hahn-Banach theorem, every point outside a convex compact set is separated from this set by a hyperplane [30]. Since the set of separable states is such a convex compact set, every entangled state can be detected by some witness. In Figure 2.6 two examples of witnesses in the bipartite state space are shown.

Optimization of Entanglement Witnesses

As mentioned before, those states for which the expectation value is negative are detected by the witness. Therefore, considering Figure 2.6 the witness W_1 detects more states than W_2 which means that W_1 is *finer* than W_2. Mathematically, this is expressed as

$$W_2 = W_1 + P, \tag{2.53}$$

where P is a positive operator. Equivalently, W_1 is finer than W_2 if and only if

$$W_2 - \alpha W_1 \geq 0 \tag{2.54}$$

holds for some $\alpha \geq 0$ [19]. If there is no finer witness than a witness W_{opt}, it is called *optimal* [9]. One necessary criterion for a witness to be optimal, is that it needs to touch the set of separable states (as the witness W_1 in Figure 2.6). This means that there has to exist a separable state σ for which $\text{Tr}(W_{\text{opt}}\sigma) = 0$. Since this criterion is not sufficient, witnesses fulfilling it are called *weakly optimal* [19].

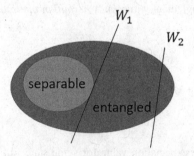

Figure 2.6 Examples of two witnesses separating the set of bipartite states. The hyperplanes are defined by $\text{Tr}(W_1 \tilde{\rho}_1) = 0$ and $\text{Tr}(W_2 \tilde{\rho}_2) = 0$ for the witnesses W_1 and W_2. The witness W_1 is *finer* than W_2

When optimizing witnesses with respect to detecting a certain state ρ, one quantity to determine how well this state is detected by the witness is the *noise robustness*.

Definition 2.7 ((White) noise robustness). *The noise robustness for a state ρ_{ent}, a witness $\mathcal{W}$ and some separable noise σ_{sep} is given by*

$$p_{crit} = \frac{Tr(\mathcal{W}\sigma_{sep})}{Tr(\mathcal{W}\sigma_{sep}) - Tr(\mathcal{W}\rho_{ent})}, \tag{2.55}$$

which is the minimal amount of ρ_{ent} needed in order to still be detected by $\mathcal{W}$. It is obtained requiring

$$Tr(\mathcal{W}\eta) = 0 \tag{2.56}$$

with $\eta = p\rho_{ent} + (1-p)\sigma_{sep}$. For the special case $\sigma_{sep} = \frac{1}{d^2}$, p_{crit} is called white noise robustness.

It is to mention that the noise robustness can also be defined as the maximum amount of noise $p'_{crit} = 1 - p_{crit}$ that can be tolerated [19]. Thus, in order to optimize a witness $\mathcal{W}$ for a certain state ρ one needs to minimize p_{crit} or maximize p'_{crit}. A picture, explaining the concept of noise robustness, is shown in Figure 2.7.

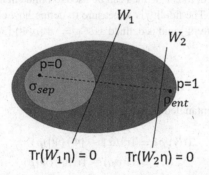

Figure 2.7 Schematic picture explaining the concept of noise robustness. Here, $p = 0$ defines the separable noise and $p = 1$ the entangled state. The dotted line shows the distance between the target state ρ_{ent} and the noise σ_{sep}, split by the witnesses $\mathcal{W}_1$ and $\mathcal{W}_2$

Construction of Entanglement Witnesses

Having defined witnesses, we now want to give a few examples how to construct them.

One way to construct optimal witnesses detecting a certain state $|\Psi\rangle$ uses the PPT criterion. In order to do so, the eigenvector $|\Phi^{\text{neg}}\rangle$ corresponding to a negative eigenvalue λ^{neg} of the partially transposed projector $|\Psi\rangle\langle\Psi|^{T_B}$ has to be computed. Then, the optimal witness detecting $|\Psi\rangle$ is given by

$$\mathcal{W}_{\text{PPT}} = |\Phi^{\text{neg}}\rangle\langle\Phi^{\text{neg}}|^{T_B}. \tag{2.57}$$

It can easily be shown that $\mathcal{W}_{\text{PPT}}$ detects the state $|\Psi\rangle$ using the property $\text{Tr}(X Y^{T_B}) = \text{Tr}(X^{T_B} Y)$ [19]:

$$\text{Tr}(\mathcal{W}_{\text{PPT}}|\Psi\rangle\langle\Psi|) = \text{Tr}\left(|\Phi^{\text{neg}}\rangle\langle\Phi^{\text{neg}}|^{T_B}|\Psi\rangle\langle\Psi|\right) \tag{2.58}$$

$$= \text{Tr}\left(|\Phi^{\text{neg}}\rangle\langle\Phi^{\text{neg}}| \, |\Psi\rangle\langle\Psi|^{T_B}\right) \tag{2.59}$$

$$= \lambda^{\text{neg}} < 0 \tag{2.60}$$

Computing the expectation value of separable states, positive eigenvalues will be obtained due to the PPT criterion. Therefore $\mathcal{W}_{\text{PPT}}$ is an entanglement witness.

Furthermore, any state close to some pure entangled state has the same properties and hence is entangled, too. This idea can be used to build entanglement witnesses based on the *fidelity*. The fidelity is a measure to define how close a general state ρ is to some pure state $|\Psi\rangle$ and is defined as $\mathcal{F}_\Psi = \langle\Psi|\rho|\Psi\rangle$ [38]. Considering the observable

$$\mathcal{W} = \alpha\mathbb{1} - |\Psi\rangle\langle\Psi|, \tag{2.61}$$

where $|\Psi\rangle$ is a pure entangled state, one finds that

$$\text{Tr}(\mathcal{W}\rho) = \text{Tr}((\alpha\mathbb{1} - |\Psi\rangle\langle\Psi|)\rho) \tag{2.62}$$

$$= \text{Tr}(\alpha\rho) - \text{Tr}(|\Psi\rangle\langle\Psi|\rho) \tag{2.63}$$

$$= \alpha - \mathcal{F}_\Psi. \tag{2.64}$$

Therefore, the observable defined in Equation (2.61) is a witness and detects any state whose fidelity exceeds a certain threshold α. To determine the minimum value for the coefficient α such that $\mathcal{W}$ is still a witness, one has to maximize the fidelity over all separable states and finds

$$\alpha = \max_{\rho \text{ separable}} \langle \Psi | \rho | \Psi \rangle \tag{2.65}$$

$$= \max_{|\Phi\rangle \text{ prod. vector}} |\langle \Phi | \Psi \rangle|^2 \tag{2.66}$$

$$= s_1^2, \tag{2.67}$$

which is the squared maximum Schmidt coefficient of $|\Psi\rangle$ [31, 38]. In the following, these types of witnesses are called *fidelity witnesses*, where in the bipartite case they are given by

$$\mathcal{W}_{\text{fid}} = s_1^2 \mathbb{1} - |\Psi\rangle\langle\Psi|. \tag{2.68}$$

Another example of a witness is found recapping the CCNR criterion, which states that for any separable state the sum of its operator Schmidt coefficients μ_i is not bigger than one. Knowing this, the *CCNR witness* can be constructed:

$$\mathcal{W}_{\text{CCNR}} = \mathbb{1} - \sum_i G_i^A \otimes G_i^B, \tag{2.69}$$

where $\{G_i^A\}$ and $\{G_i^B\}$ are the orthonormal operators obtained from the operator Schmidt decomposition. This witness detects any state ρ_{ent} violating the CCNR criterion, since, using the orthonormality of the $\{G_i^A\}$ and $\{G_i^B\}$, it is

$$\text{Tr}(\mathcal{W}_{\text{CCNR}}\rho_{\text{ent}}) = 1 - \sum_i \mu_i < 0. \tag{2.70}$$

Furthermore, its expectation value is positive for all separable states, as for any state $\rho = \sum_{ij} \alpha_{ij} G_i^A \otimes G_j^B$ it is

$$\text{Tr}(\mathcal{W}_{\text{CCNR}}\rho) = 1 - \sum_i \alpha_{ii} = 1 - \text{Tr}(A). \tag{2.71}$$

Using that the trace of a matrix is always smaller than the sum of its singular values (and remembering that the operator Schmidt coefficients are nothing else but singular values), we find

$$\text{Tr}(\mathcal{W}_{\text{CCNR}}\rho) \geq 1 - \sum_i \mu_i \geq 0 \tag{2.72}$$

since $\sum_i \mu_i \leq 1$ for separable states [19].

2.3.2 Schmidt Witnesses

Previously, the concept of the Schmidt number was introduced. In order to determine the Schmidt number of a state one can use *Schmidt witnesses*. These are defined in the same way as the entanglement witnesses in Subsection 2.3.1.

As already known, states with Schmidt number k form a convex set in the set of all bipartite states. These subsets are denoted S_k and fulfill $S_{k-1} \subset S_k$. A *k-Schmidt witness* detects states with Schmidt number k or higher and is defined as follows [7]:

Definition 2.8 (Schmidt witness). *A hermitian operator $\mathcal{W}_{kS}$ is a k-Schmidt witness if and only if it fulfills*

$$Tr(\mathcal{W}_{kS}\sigma) \geq 0 \ \forall \ \sigma \in S_{k-1} \tag{2.73}$$

and there exists at least one state $\rho \in S_k$ such that

$$Tr(\mathcal{W}_{kS}\rho) < 0. \tag{2.74}$$

Since the set S_1 is equal to the set of separable states, any 2-Schmidt witness is equivalent to an entanglement witness detecting inseparable states as in the previous section. Moreover, every k-Schmidt witness can be written in the canonical form

$$\mathcal{W} = W - \varepsilon \mathbb{1}, \tag{2.75}$$

where W is a positive operator, which has no vectors of S_{k-1} in its kernel, i. e. there exists no vector $|\Psi^{<k}\rangle \in S_{k-1}$ such that $W|\Psi^{<k}\rangle = 0$. In addition, ε fulfills $0 < \varepsilon \leq \inf_{|\Psi^{<k}\rangle \in S_{k-1}} \langle \Psi^{<k}|W|\Psi^{<k}\rangle$ [7].

In Figure 2.8 an example of a 3-Schmidt witness is shown.

Considering the figure, one can see that the hyperplane touches the set S_2. Then, the corresponding witness is called *tangent*. Mathematically, this means that for any k-Schmidt witness $\mathcal{W}_{kS}$ it is tangent if and only if there exists a state $\rho \in S_{k-1}$ such that

$$Tr(\mathcal{W}_{kS}\rho) = 0. \tag{2.76}$$

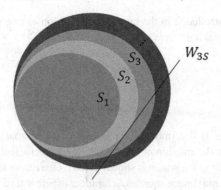

Figure 2.8 Example of a 3-Schmidt witness. The hyperplane separates the bipartite state space in the part of the detected states with Schmidt number $k = 3$ or higher and the part of states with positive expectation value, including all of those with Schmidt number two or lower. Furthermore, one can see that the hyperplane $\mathcal{W}_{3S}$ touches the set S_2

With this definition one can find that a state ρ is in S_{k-1} if and only if

$$\mathrm{Tr}(\mathcal{W}_{kS}\rho) \geq 0 \tag{2.77}$$

for all $\mathcal{W}_{kS}$ that are tangent to S_{k-1} [7].

Lastly, we want to give a more specific example: As well as entanglement witnesses, one can also define Schmidt number witnesses using the fidelity. For an arbitrary dimension d and Schmidt number k, these witnesses are of the form [7]:

$$\mathcal{W}_k = \frac{k-1}{d}\mathbb{1} - |\Psi_+\rangle\langle\Psi_+|, \tag{2.78}$$

where $|\Psi_+\rangle = \frac{1}{\sqrt{d}}\sum_{i=0}^{d-1}|ii\rangle$ is a maximally entangled state.

2.3.3 Multipartite Entanglement Witnesses

As mentioned before, entanglement witnesses can also be defined for the multipartite case. An observable with positive expectation values on $(N-1)$-partite entangled states and negative expectation value for at least one N-partite entangled state is then a *witness of genuine N-partite entanglement* [31]. The fidelity witness

(Equation (2.61)) introduced in the previous subsection can be generalized to the multipartite case by taking

$$\alpha = \max_{|\Phi\rangle \in B} |\langle\Phi|\Psi\rangle|^2 \tag{2.79}$$

$$= \max_{\text{all bipartitions}} s_1^2, \tag{2.80}$$

where the maximum is determined over the set of all biseparable states B. The parameter α is then the maximum squared Schmidt coefficient over all possible bipartitions of the state $|\Psi\rangle$, which is shown in [31]. One example is the witness for the GHZ state. The maximum squared Schmidt coefficient taken over all possible bipartitions is $\alpha = \frac{1}{2}$, which can be seen directly from the definition of $|GHZ\rangle$ (Equation (2.46) in Section 2.2.2). The corresponding fidelity witness is then given by

$$\mathcal{W} = \frac{1}{2}\mathbb{1} - |GHZ\rangle\langle GHZ|. \tag{2.81}$$

However, since there are several classes of GME, witnesses detecting states in those classes can be defined, too. As an example, entanglement witnesses for the tripartite case are discussed in the following.

For three qubits there are two classes of genuine tripartite entanglement, the GHZ class and the W class, therefore GHZ and W witnesses can be defined. A GHZ witness $\mathcal{W}_{\text{GHZ}}$ needs to have positive expectation values for all states $\rho_W \in W$ from the W class and a negative expectation value for at least one $\rho \in GHZ \setminus W$, which is in the GHZ class but can not be expressed as a convex sum of W type, biseparable and product vectors. A W witness $\mathcal{W}_W$ needs to fulfill $\text{Tr}(\mathcal{W}_W \rho_B) \geq 0$ for all biseparable states $\rho_B \in B$ and $\text{Tr}(\mathcal{W}_W \rho_W) < 0$ for at least one state $\rho_W \in W \setminus B$ [30]. Any of those witnesses can be expressed in the form

$$\mathcal{W} = Q - \varepsilon\mathbb{1}. \tag{2.82}$$

The operator Q is positive and has no W type (or GHZ type) vectors in its kernel, which means that there is no W type (GHZ type) vector $|\phi\rangle$ such that $Q|\phi\rangle = 0$. A witness fulfilling this for the GHZ class, is given by

$$\mathcal{W}_{\text{GHZ}} = \frac{3}{4}\mathbb{1} - |GHZ\rangle\langle GHZ|. \tag{2.83}$$

This is again a fidelity witness, where $\alpha = \frac{3}{4} = \max\limits_{|\Psi_W\rangle \in W} |\langle GHZ|\Psi_W\rangle|^2$ is the maximum squared overlap between $|GHZ\rangle$ and a W type vector.

For the W class it is found, for example,

$$\mathcal{W}_W = \frac{2}{3}\mathbb{1} - |W\rangle\langle W|, \tag{2.84}$$

where $\alpha = \frac{2}{3}$ is the maximum squared overlap between $|W\rangle$ and a B type vector $|\Psi_B\rangle$ [30].

2.4   Applications of Quantum Entanglement

While in the previous parts the concept of entanglement was introduced and mathematically formulated, now we want to make clear why entanglement is useful. At the beginning there is to mention that entanglement was first noticed by Einstein, Podolsky and Rosen and Schrödinger in 1935 as "spooky" feature of quantum mechanics [24]. Nowadays this phenomenon is of large interest in quantum information theory as it is a resource for many applications, such as *quantum teleportation*, *quantum cryptography* and *quantum metrology*. In the following, those three examples will be discussed.

2.4.1   Quantum Teleportation

Imagine the situation where Alice and Bob share an entangled quantum state. Now, Alice wants to transmit an unknown state $|\psi\rangle$ to Bob using only classical communication. This seems to be impossible, since even if Alice knew the state it would take her infinitely long to give all information to Bob as the state takes values in a continuous space. However, quantum teleportation makes this task possible [42]. In order to *teleport* the qubit state $|\psi\rangle$ to Bob, first, Alice and Bob have to create a *maximally entangled* state

$$|\Psi^-\rangle_{AB} = \frac{1}{\sqrt{2}}(|0\rangle_A|1\rangle_B - |1\rangle_A|0\rangle_B). \tag{2.85}$$

This state is called *EPR (Einstein, Podolsky, Rosen) singlet state* (or later simply *singlet state*) and is one of the four *Bell states*:

$$|\Psi^{\pm}\rangle = \frac{1}{\sqrt{2}}(|01\rangle \pm |10\rangle) \tag{2.86}$$

$$|\Phi^{\pm}\rangle = \frac{1}{\sqrt{2}}(|00\rangle \pm |11\rangle). \tag{2.87}$$

Now, Alice and Bob share the EPR state (2.85), where Alice obtains the qubits denoted by A and Bob those denoted by B. In total, the state is then $|\psi\rangle_S \otimes |\Psi^-\rangle_{AB}$, which is a pure product state and therefore, there is no entanglement between the state $|\psi\rangle$ Alice wants to send and the shared EPR pair [46]. In order to create entanglement, Alice now performs a von Neumann measurement in the Bell operator basis, consisting of the states in equations (2.86) and (2.87) on her part of the system. Before the measurement, the state can be expressed in the following form

$$|\Psi\rangle_{SAB} = \frac{1}{2}(|\Psi^-\rangle_{SA}(-a|0\rangle_B - b|1\rangle_B) \tag{2.88}$$

$$+|\Psi^+\rangle_{SA}(-a|0\rangle_B + b|1\rangle_B) \tag{2.89}$$

$$+|\Phi^-\rangle_{SA}(a|1\rangle_B + b|0\rangle_B) \tag{2.90}$$

$$+|\Phi^+\rangle_{SA}(a|1\rangle_B - b|0\rangle_B)). \tag{2.91}$$

The coefficients a and b appear since the unknown state $|\psi\rangle$ may be written as

$$|\psi\rangle = a|0\rangle_S + b|1\rangle_S \tag{2.92}$$

with $|a|^2 + |b|^2 = 1$. Thus, after the measurement Bobs state is projected to one of the pure states in equation lines (2.88) to (2.91) depending on the outcome [46]. Since Bob then already has the state $|\psi\rangle$ up to a unitary transformation, namely

$$-|\psi\rangle_B = -\begin{pmatrix} 1 & 0 \\ 0 & 1 \end{pmatrix} |\psi\rangle_B \tag{2.93}$$

$$-|\psi\rangle_B = \begin{pmatrix} -1 & 0 \\ 0 & 1 \end{pmatrix} |\psi\rangle_B \tag{2.94}$$

$$-|\psi\rangle_B = \begin{pmatrix} 0 & 1 \\ 1 & 0 \end{pmatrix} |\psi\rangle_B \tag{2.95}$$

$$-|\psi\rangle_B = \begin{pmatrix} 0 & -1 \\ 1 & 0 \end{pmatrix} |\psi\rangle_B, \tag{2.96}$$

Alice just has to tell her measurement result to Bob. Then Bob can apply the correct unitary according to equations (2.93) to (2.96) and ends up with the desired state

$|\psi\rangle$. Alice's state is then one of the Bell states, which means her state $|\psi\rangle$ was teleported to Bob. Figure 2.9 shows the scheme of teleportation.

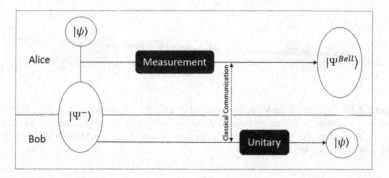

Figure 2.9 Teleportation scheme. The entangled states are drawn in egg shape, where $|\Psi^-\rangle$ is the singlet state and $|\Psi^{Bell}\rangle$ denotes one of the Bell states. The circled state $|\psi\rangle$ is the one Alice teleports to Bob

Note that this procedure can also be generalized to mixed states and systems having more than two orthogonal states, which means that Alice would use a completely entangled pair of N-state particles instead of the EPR pair [46].

2.4.2   Quantum Cryptography

Another important application in quantum information is *quantum cryptography* or *quantum key distribution (QKD)*. This time, Alice and Bob want to communicate securely, which means that for an eavesdropper Eve it should not be possible to get their messages without Alice and Bob noticing. Two principles to reach this are *private* and *public key cryptography*. The following example will be reduced to the principle of private key cryptography, which is provably secure.

Private Key Cryptography and The Vernam Cipher

First, Alice needs to *encode* her message with a private key. After that she can send the encrypted message to Bob using a *public channel*. Bob then uses the key to *decode* the message. One simple but very effective cryptosystem is the *Vernam cipher*. The idea is that Alice and Bob both have identical n-bit secret key strings. To encode her message, Alice needs to add the key string to it, where Bob can decode the message by subtracting the key string from the received message. However, this can only be secure if the key is at least as large as the message itself. The scheme of the Vernam cipher is shown in Figure 2.10 [42].

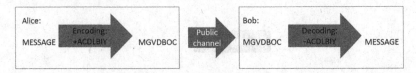

Figure 2.10 Scheme of the Vernam cipher. Alice encodes her message by adding the key, where the letters in the key assign how much the corresponding letter in the message has to be shifted. Then, she sends the encrypted message through a public channel to Bob who encrypts it by subtracting the key

The Ekert Protocol

Clearly, the security of the communication between Alice and Bob depends on the fact that only the two of them know the encoding/decoding key. Thus, the task of quantum key distribution is to find *protocols* that generate such keys with maximum privacy. Those protocols can be separated in two classes, the *prepare and measure protocols (P&M)* and the *entanglement-based protocols*. One example of an entanglement-based protocol is the *Ekert protocol* which was introduced by Athur K. Ekert in 1991 and therefore is also called *E91* protocol [12].

The idea is the following: Having a source producing singlet states $|\Psi^-\rangle$, Alice and Bob perform measurements along one of three directions $\vec{a}_i$ and $\vec{b}_j$. They choose randomly between the azimuthal angles

$$\phi_1^a = 0, \ \phi_2^a = \frac{\pi}{4}, \ \phi_3^a = \frac{\pi}{2} \tag{2.97}$$

$$\phi_1^b = \frac{\pi}{4}, \ \phi_2^b = \frac{\pi}{2}, \ \phi_3^b = \frac{3\pi}{4} \tag{2.98}$$

and may obtain results $+1$ or -1, where the upper index of ϕ_i^α denotes if the angle corresponds to Alice or Bob and the lower index fixes the measurement direction. The correlation coefficient of the measurements performed by Alice and Bob is given by

$$E(\vec{a}_i, \vec{b}_j) = P_{++}(\vec{a}_i, \vec{b}_j) + P_{--}(\vec{a}_i, \vec{b}_j) - P_{+-}(\vec{a}_i, \vec{b}_j) - P_{-+}(\vec{a}_i, \vec{b}_j), \tag{2.99}$$

where $P_{\pm\pm}(\vec{a}_i, \vec{b}_j)$ is the probability that Alice, having measured $\vec{a}_i$, obtained ± 1 and Bob, having measured $\vec{b}_j$, obtained ± 1. Due to the laws of quantum mechanics, if Alice and Bob measure the singlet state in the same direction (which would be the case for $(\vec{a}_2, \vec{b}_1)$ and $(\vec{a}_3, \vec{b}_2)$), the results will be completely anticorrelated, which means $E(\vec{a}_2, \vec{b}_1) = E(\vec{a}_3, \vec{b}_2) = -1$.

When Alice and Bob now have done their measurements, they can compare their chosen directions $\vec{a}_i$ and $\vec{b}_j$ in public and split their results into two groups. In the first group they put all results where they had chosen the same direction. To the second group belong all results where they had chosen different directions. Due to the perfect anticorrelation, they can use the measurement results of the first group to establish their key. The results from the second group they can use to check if an eavesdropper had intercepted some information. For this, the following quantity has to be considered:

$$S = E(\vec{a}_1, \vec{b}_1) - E(\vec{a}_1, \vec{b}_3) + E(\vec{a}_3, \vec{b}_1) + E(\vec{a}_3, \vec{b}_3). \qquad (2.100)$$

If Alice and Bob really share the state $|\Psi^-\rangle$, S is required to fulfill

$$S = -2\sqrt{2}. \qquad (2.101)$$

Thus, Alice and Bob can compute S from their results in group two and if they obtain $S \neq -2\sqrt{2}$, they can be sure that there was an eavesdropper involved.

There is to mention that the source, producing the singlet state, might be Eve herself. However, for one thing, if she would send another state than $|\Psi^-\rangle$ to Alice and Bob, Equation (2.101) would not hold and hence her disturbance would be detected. For another thing, if she would try to create a mixed three-qubit state ρ_{ABE} in order to have correlation between her measurements and the measurements of Alice and Bob, this is not possible, since Alice and Bob hold the maximally entangled state $|\Psi^-\rangle\langle\Psi^-|$ and hence there can not exist further entanglement between Eve and Alice and Bob without disturbing their pure state.

Entanglement Witnesses in QKD

As mentioned before, in any protocol, having established a key, Alice and Bob want to be sure that it is secure. In the E91 protocol this was achieved by checking Equation (2.101). However, entanglement witnesses give another tool to proof that a key might not be secure.

Every QKD protocol consists of two phases, where in the first phase, Alice and Bob generate correlated data, described by $P(A, B)$. In [32] it was proven that this data can only be used to establish a secret key if there are quantum correlations in

$P(A, B)$. This means that if $P(A, B)$ can be interpreted as coming from a separable state, it is not possible to establish a secret key from this data via public communication. Since the occurrence of entanglement is therefore a necessary condition, it needs to be detected. Here, entanglement witnesses come into account as they were shown to give necessary and sufficient conditions for separability even if the state can not be completely reconstructed [32, 33]. For the entanglement-based protocols this can be formulated in the following theorem [32]:

Theorem 2.3 (Entanglement witnesses in QKD). *Let Alice and Bob perform local measurements with POVM elements $A_i \otimes B_i$ on the distributed state ρ and obtain $P(A, B)$. Then $P(A, B)$ can not origin from a separable state if and only if there is an entanglement witness $\mathcal{W} = \sum_i c_i A_i \otimes B_i$ detecting ρ.*

Thus, if the state ρ is not detected by the witness $\mathcal{W}$, one may be sure that $P(A, B)$ originates from a separable state and hence no secret key can be created.

2.4.3   Quantum Metrology

When doing measurements in order to estimate certain parameters, quantum effects like squeezing and entanglement can be used to increase the accuracy of the estimation; these strategies are part of *quantum metrology* [35, 36]. More precisely, when doing measurements the laws of quantum mechanics, for example the Heisenberg uncertainty relation limit the accuracy. Unlike this inaccuracies, there are other sources of uncertainty, e. g. environment induced noise from vacuum fluctuations, which can be decreased by choosing an optimal measurement strategy. Thus, some tools of quantum mechanics can be used to our advantage and make the measurement results more precise. In the following, this will be shown.

Assuming there is a setup, such that a qubit state $|\psi_{\text{in}}\rangle = \frac{1}{\sqrt{2}}(|0\rangle + |1\rangle)$ is transformed into the state $|\psi_{\text{out}}\rangle = \frac{1}{\sqrt{2}}(|0\rangle + e^{i\varphi}|1\rangle)$. The goal is to estimate the phase φ as exact as possible [36]. Experimentally, it can be checked when the output state is equal to the input state, since the probability for this is

$$p(\varphi) = |\langle \psi_{\text{in}}|\psi_{\text{out}}\rangle|^2 = \cos^2(\varphi/2) \tag{2.102}$$

and thus the phase φ can be determined. Using methods of statistics like error propagation, the uncertainty of φ is

$$\Delta\varphi = \frac{\Delta p(\varphi)}{|\frac{\partial p(\varphi)}{\partial\varphi}|} = 1, \tag{2.103}$$

where $\Delta^2 p(\varphi)$ is the variance of $p(\varphi)$ and the result 1 is obtained only in this case by inserting the function of $p(\varphi)$ (Eq. (2.102)). In order to improve this estimation, the experiment can be repeated N times. Thus, using the central limit theorem, the error is

$$\Delta\varphi \propto \frac{1}{\sqrt{N}}, \tag{2.104}$$

which is known as the *standard quantum limit*. To beat this limit, entanglement can be used. Therefore, instead of measuring N single qubits, the following state can be prepared

$$|\Psi_{\text{in}}\rangle = \frac{1}{\sqrt{2}}(|0\rangle...|0\rangle + |1\rangle...|1\rangle). \tag{2.105}$$

This is an entangled N-qubit state and the output, measured by a non-local measurement, then looks similar to the one-qubit case:

$$|\Psi_{\text{out}}\rangle = \frac{1}{\sqrt{2}}(|0\rangle...|0\rangle + e^{iN\varphi}|1\rangle...|1\rangle). \tag{2.106}$$

Going through the same computations as before, the uncertainty is found to be

$$\Delta\varphi \propto \frac{1}{N}, \tag{2.107}$$

which is an $\sqrt{N}$ enhancement with respect to the standard quantum limit in Equation (2.104).

There are two things left to mention. For one thing, in [35] four different strategies where investigated, where the input state was either prepared classically (C) or entangled (Q) as well as the measurements, which were either performed separable (C) or non-local (Q). It was shown that in order to beat the quantum limit, the preparation of an entangled state is required whereas measurements can be done separably. A scheme of this strategy, compared to the completely classical one, is shown in Figure 2.11. Moreover, for the other thing, it was proven that the new precision $\propto \frac{1}{N}$ is optimal.

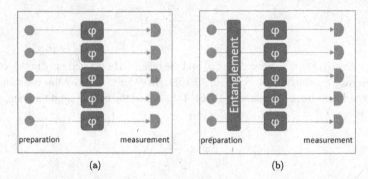

(a)   (b)

Figure 2.11 Measurement scheme of a classical and a quantum strategy. In the classical strategy (a) N qubits are prepared, obtain a phase shift φ and are measured locally. In the quantum ansatz (b), the input state is entangled and separable measurements are performed

OSD Witnesses for Bipartite States

In this chapter we introduce the new type of witness, which is based on the Schmidt decomposition of operators. In the first part we will give the main idea and show how to construct the entanglement witness. Further, we will show that it is strictly stronger than fidelity-based witnesses. In the second part some approaches to numerically find those witnesses are discussed.

3.1 Theoretical Considerations

As mentioned before, first, the main idea of the new concept will be discussed.

3.1.1 Construction for the Bipartite Case

The previously introduced fidelity witness, which is based on the Schmidt decomposition of vectors can be generalized using the Schmidt decomposition of operators. It is given by the following [3].

Supplementary Information The online version contains supplementary material available at https://doi.org/10.1007/978-3-658-43203-4_3.

S. Denker, *Characterizing Multiparticle Entanglement Using the Schmidt Decomposition of Operators*, BestMasters, https://doi.org/10.1007/978-3-658-43203-4_3

Theorem 3.1 (Bipartite OSD witnesses). *Let* $X = \sum_i \mu_i G_i^A \otimes G_i^B$ *be a hermitian operator in its operator Schmidt decomposition. Then*

$$\mathcal{W}_{\text{OSD}} = \mu_1 \mathbb{1} - X \tag{3.1}$$

is an entanglement witness.

In order to let an observable $\mathcal{W} = \lambda \mathbb{1} - X$ be an entanglement witness, it has to fulfill $\text{Tr}(\mathcal{W}\rho_{\text{sep}}) \geq 0$ for all separable states ρ_{sep}. That means:

$$0 \overset{!}{\leq} \text{Tr}(\mathcal{W}\rho_{\text{sep}}) \tag{3.2}$$
$$= \text{Tr}((\lambda \mathbb{1} - X)\rho_{\text{sep}}) \tag{3.3}$$
$$= \lambda\text{Tr}(\rho_{\text{sep}}) - \text{Tr}(X\rho_{\text{sep}}) \tag{3.4}$$
$$\Rightarrow \lambda \overset{!}{\geq} \text{Tr}(X\rho_{\text{sep}}) \tag{3.5}$$

In the following proof we show that Condition (3.5) is fulfilled by $\lambda = \mu_1$ and therefore $\mathcal{W}_{\text{OSD}}$ is positive on all separable states.

Proof. We start by considering $\text{Tr}(X\rho_{\text{sep}})$ and expressing X and ρ_{sep} in the operator Schmidt decomposition:

$$\text{Tr}(X\rho_{\text{sep}}) = \text{Tr}\left(\sum_i \mu_i G_i^A \otimes G_i^B \sum_j \gamma_j H_j^A \otimes H_j^B\right) \tag{3.6}$$
$$= \sum_{ij} \mu_i \gamma_j \text{Tr}(G_i^A H_j^A)\text{Tr}(G_i^B H_j^B). \tag{3.7}$$

In this step we used the relation $\text{Tr}(A \otimes B) = \text{Tr}(A)\text{Tr}(B)$ as well as the linearity of the trace. Furthermore, a basis transformation $G_i^{A/B} = \sum_k O_{ik}^{A/B} H_k^{A/B}$ can be done and it follows:

$$\text{Tr}(X\rho_{\text{sep}}) = \sum_{ij} \mu_i \gamma_j \text{Tr}\left(\sum_k O_{ik}^A H_k^A H_j^A\right) \text{Tr}\left(\sum_k O_{ik}^B H_k^B H_j^B\right) \tag{3.8}$$

$$= \sum_{ij} \left\{ \mu_i \gamma_j \left(\sum_k O_{ik}^A \text{Tr}(H_k^A H_j^A)\right) \left(\sum_k O_{ik}^B \text{Tr}(H_k^B H_j^B)\right) \right\} \tag{3.9}$$

$$= \sum_{ij} \left\{ \mu_i \gamma_j \left(\sum_k O_{ik}^A \delta_{kj}\right) \left(\sum_k O_{ik}^B \delta_{kj}\right) \right\} \tag{3.10}$$

$$= \sum_{ij} \mu_i \gamma_j O_{ij}^A O_{ij}^B \tag{3.11}$$

$$= \sum_{ij} \sqrt{\mu_i \gamma_j} O_{ij}^A \sqrt{\mu_i \gamma_j} O_{ij}^B. \tag{3.12}$$

In equation line (3.10) we used that the Schmidt operators are hermitian and form an orthonormal basis, which means that $\text{Tr}\left(H_i^{A/B} H_j^{A/B}\right) = \text{Tr}\left((H_i^{A/B})^\dagger H_j^{A/B}\right) = \delta_{ij}$. The expression in line (3.12) can be interpreted as a scalar product and therefore we can apply the Cauchy-Schwarz inequality. Moreover, the scalar product of two vectors a and b gets maximal if $a = b$.

$$\text{Tr}(X\rho_{\text{sep}}) \leq \sum_{ij} |\sqrt{\mu_i \gamma_j} O_{ij}^A| |\sqrt{\mu_i \gamma_j} O_{ij}^B| \tag{3.13}$$

$$\leq \sum_{ij} \mu_i \gamma_j |O_{ij}^A|^2. \tag{3.14}$$

The term $|O_{ij}^A|^2$ in equation line (3.14) describes the entries of a double stochastic matrix, i.e. a matrix whose columns and rows sum up to one. Thus, Birkhoff's theorem can be applied, which states that every double stochastic matrix can be written as convex combination of permutation matrices yielding

$$\text{Tr}(X\rho_{\text{sep}}) \leq \sum_i \mu_i \gamma_{\pi(i)}, \tag{3.15}$$

where the coefficients $\gamma_{\pi(i)}$ are permuted. Therefore, rearranging the coefficients, such that they are sorted decreasingly (denoted by $\downarrow$), makes the term larger again:

$$\text{Tr}(X\rho_{\text{sep}}) \leq \sum_i \mu_i \gamma_{i\downarrow}. \tag{3.16}$$

Finally, we make use of the fact that convex functions attain their maximum within the extreme points of the set, which for the set of separable states are the pure product states: $\rho_{PS} = \rho_A \otimes \rho_B$. These states are already expressed in the operator Schmidt decomposition and have only one nonzero operator Schmidt coefficient $\gamma_1 = 1$. It follows:

$$\text{Tr}(X\rho_{\text{sep}}) \leq \mu_1 \qquad (3.17)$$

and thus Condition (3.5) is fulfilled by $\lambda = \mu_1$.   □

3.1.2   Comparison with the Fidelity Witness

In the previous subsection it was mentioned that the OSD witness is a generalization of the fidelity witness. Now we will see why this is the case.

First, we consider a pure state $|\Psi\rangle = \sum_k s_k |\alpha_k \beta_k\rangle$ in its vector Schmidt decomposition and form the projector $|\Psi\rangle\langle\Psi|$. Then, the Schmidt decomposition in operator space is performed:

$$|\Psi\rangle\langle\Psi| = \sum_{kl} s_k s_l |\alpha_k \beta_k\rangle\langle\alpha_l \beta_l| \qquad (3.18)$$

$$= \sum_{k=l} s_k^2 |\alpha_k \beta_k\rangle\langle\alpha_k \beta_k| + \sum_{k\neq l} s_k s_l |\alpha_k \beta_k\rangle\langle\alpha_l \beta_l| \qquad (3.19)$$

$$= \sum_{k=l} s_k^2 |\alpha_k\rangle\langle\alpha_k| \otimes |\beta_k\rangle\langle\beta_k| + \sum_{k\neq l} s_k s_l |\alpha_k\rangle\langle\alpha_l| \otimes |\beta_k\rangle\langle\beta_l| \qquad (3.20)$$

$$= \sum_i \mu_i M_i^A \otimes M_i^B. \qquad (3.21)$$

The $M_i^{A/B}$ fulfill $\text{Tr}(M_i^\dagger M_j) = \delta_{ij}$ but they are not hermitian, yet. However, taking a closer look at the terms of the second sum in equation line (3.20), one finds that the matrices in the terms $s_k s_l (|\alpha_k\rangle\langle\alpha_l| \otimes |\beta_k\rangle\langle\beta_l| + |\alpha_l\rangle\langle\alpha_k| \otimes |\beta_l\rangle\langle\beta_k|)$ can be interpreted as Pauli matrices and therefore are hermitian. Thus, the operator Schmidt coefficients of a pure state $|\Psi\rangle$ are given by the products of its vector Schmidt coefficients:

$$\{\mu_i\} = \{s_k s_l\}. \qquad (3.22)$$

This means that the bipartite fidelity witness (Equation (2.68)) is simply a special case of the OSD witness where $X = |\Psi\rangle\langle\Psi|$, since the maximum operator Schmidt coefficient $\mu_1 = s_1^2$ is given by the squared maximum vector Schmidt coefficient.

Furthermore, one can take a look at the CCNR witness, which was introduced in Subsection 2.3.1:

$$\mathcal{W}_{\text{CCNR}} = \mathbb{1} - \sum_i G_i^A \otimes G_i^B. \tag{3.23}$$

Clearly, this is also a special case of the OSD witness, where the operator X has only operator Schmidt coefficients equal to one and zero.

Those two observations lead to the following: Using results from [47], in [38] it was shown that only entangled states that violate the PPT criterion can be detected by fidelity witnesses. Moreover, the CCNR witness detects those states violating the CCNR criterion. As mentioned in Subsection 2.2.1 the PPT and the CCNR criterion complement each other. Thus, the OSD witness can detect entangled states that violate the CCNR criterion as well as the PPT criterion and therefore is strictly stronger than the fidelity witness, which may only detect NPT states.

3.2 Optimization

As noted in the previous sections, the OSD witness is given by

$$\mathcal{W}_{\text{OSD}} = \mu_1 \mathbb{1} - X, \tag{3.24}$$

where μ_1 is the largest operator Schmidt coefficient of the operator X. However, it was not mentioned yet how to choose X. This will be discussed in the following. The first part introduces an algorithm, optimizing with respect to the operator Schmidt coefficients and in the second part the optimization is focused on the Schmidt operators. Lastly, we will discuss some examples for the practical use.

3.2.1 Optimization with Respect to The OS Coefficients

In order to optimize the OSD witness, the goal is to minimize the (white) noise robustness (Definition 2.7) for a given pure state $\rho_{\text{ent}} = |\Psi\rangle\langle\Psi|$, that is aimed to be detected by the witness:

$$p_{\text{crit}} = \frac{\text{Tr}(\mathcal{W}_{\text{OSD}}\sigma_{\text{sep}})}{\text{Tr}(\mathcal{W}_{\text{OSD}}\sigma_{\text{sep}}) - \text{Tr}(\mathcal{W}_{\text{OSD}}\rho_{\text{ent}})}. \tag{3.25}$$

This equation can be rewritten using the definition of the OSD witness and expressing the operator X in its operator Schmidt decomposition:

$$p_{\text{crit}} = \frac{\text{Tr}((\mu_1\mathbb{1} - X)\sigma_{\text{sep}})}{\text{Tr}((\mu_1\mathbb{1} - X)\sigma_{\text{sep}}) - \text{Tr}((\mu_1\mathbb{1} - X)\rho_{\text{ent}})} \tag{3.26}$$

$$= \frac{\mu_1 - \text{Tr}(X\sigma_{\text{sep}})}{\mu_1 - \text{Tr}(X\sigma_{\text{sep}}) - (\mu_1 - \text{Tr}(X\rho_{\text{ent}}))} \tag{3.27}$$

$$= \frac{\mu_1 - \text{Tr}(X\sigma_{\text{sep}})}{\text{Tr}(X\rho_{\text{ent}}) - \text{Tr}(X\sigma_{\text{sep}})} \tag{3.28}$$

$$= \frac{\mu_1 - \text{Tr}\left(\left(\sum_i \mu_i G_i^A \otimes G_i^B\right)\sigma_{\text{sep}}\right)}{\text{Tr}\left(\left(\sum_i \mu_i G_i^A \otimes G_i^B\right)\rho_{\text{ent}}\right) - \text{Tr}\left(\left(\sum_i \mu_i G_i^A \otimes G_i^B\right)\sigma_{\text{sep}}\right)} \tag{3.29}$$

$$= \frac{\mu_1 - \sum_i \mu_i\text{Tr}((G_i^A \otimes G_i^B)\sigma_{\text{sep}})}{\sum_i \mu_i\text{Tr}((G_i^A \otimes G_i^B)\rho_{\text{ent}}) - \sum_i \mu_i\text{Tr}((G_i^A \otimes G_i^B)\sigma_{\text{sep}})} \tag{3.30}$$

$$= \frac{\mu_1 - \sum_i \mu_i\text{Tr}((G_i^A \otimes G_i^B)\sigma_{\text{sep}})}{\sum_i \mu_i\text{Tr}((G_i^A \otimes G_i^B)(\rho_{\text{ent}} - \sigma_{\text{sep}}))}. \tag{3.31}$$

Finally, equation line (3.31) can be interpreted as a function of the operator Schmidt coefficients μ_i:

$$p_{\text{crit}} = p_{\text{crit}}(\vec{\mu}), \tag{3.32}$$

with $\vec{\mu} = (\mu_1, \mu_2, ...)$. Since the goal is to minimize the (white) noise robustness, a *gradient descent algorithm* (see Appendix A in the Electronic Supplementary Material) can be used. For this, one has to compute the gradient of $p_{\text{crit}}(\vec{\mu})$ and to do so, we first define:

$$\tilde{\rho}_i := \text{Tr}((G_i^A \otimes G_i^B)\rho_{\text{ent}}) \tag{3.33}$$

$$\tilde{\sigma}_i := \text{Tr}((G_i^A \otimes G_i^B)\sigma_{\text{sep}}). \tag{3.34}$$

Then the (white) noise robustness can be expressed in a simpler way:

$$p_{\text{crit}} = \frac{\mu_1 - \sum_i \mu_i\tilde{\sigma}_i}{\sum_i \mu_i(\tilde{\rho}_i - \tilde{\sigma}_i)}. \tag{3.35}$$

To compute the gradient, we start with the derivation with respect to μ_1. Using the quotient rule for derivations, this yields

$$\frac{dp_{\text{crit}}}{d\mu_1} = \frac{(1 - \tilde{\sigma}_1)\left(\sum_i \mu_i (\tilde{\rho}_i - \tilde{\sigma}_i)\right) - (\tilde{\rho}_1 - \tilde{\sigma}_1)\left(\mu_1 - \sum_i \mu_i \tilde{\sigma}_i\right)}{\left(\sum_i \mu_i (\tilde{\rho}_i - \tilde{\sigma}_i)\right)^2} \tag{3.36}$$

and analogously the derivation with respect to μ_j for $j \neq 1$ can be computed:

$$\frac{dp_{\text{crit}}}{d\mu_j} = \frac{(-\tilde{\sigma}_j)\left(\sum_i \mu_i (\tilde{\rho}_i - \tilde{\sigma}_i)\right) - (\tilde{\rho}_j - \tilde{\sigma}_j)\left(\mu_1 - \sum_i \mu_i \tilde{\sigma}_i\right)}{\left(\sum_i \mu_i (\tilde{\rho}_i - \tilde{\sigma}_i)\right)^2}. \tag{3.37}$$

Thus, the j-th entry of the whole gradient is given by

$$(\vec{\nabla}_{\vec{\mu}} p_{\text{crit}}(\vec{\mu}))_j = \frac{(\delta_{1j} - \tilde{\sigma}_j)\left(\sum_i \mu_i (\tilde{\rho}_i - \tilde{\sigma}_i)\right) - (\tilde{\rho}_j - \tilde{\sigma}_j)\left(\mu_1 - \sum_i \mu_i \tilde{\sigma}_i\right)}{\left(\sum_i \mu_i (\tilde{\rho}_i - \tilde{\sigma}_i)\right)^2}. \tag{3.38}$$

With that, the (white) noise robustness can be optimized using the following algorithm:

Algorithm 3.1 (OSC optimization for two parties).

- *Start with some input operator X_{in} and compute the OSD witness*
- *Write $X_{\text{in}} = \sum_i \mu_i G_i^A \otimes G_i^B$ in the operator Schmidt decomposition*
- *Compute the gradient $\vec{\nabla}_{\vec{\mu}} p_{\text{crit}}(\vec{\mu})$ of the white noise robustness ($\sigma_{\text{sep}} = \frac{1}{d^2}$)*
- *Update the operator Schmidt coefficients according to the update rule*

$$\vec{\mu} \mapsto \vec{\mu} - \varepsilon \vec{v} \text{ with } \vec{v} = \frac{\vec{\nabla}_{\vec{\mu}} p_{\text{crit}}(\vec{\mu})}{\|\vec{\nabla}_{\vec{\mu}} p_{\text{crit}}(\vec{\mu})\|} \tag{3.39}$$

- *Update the operator X_{in} using the new operator Schmidt coefficients:*

$$X_{\text{up}} = \sum_i \mu_{i,\text{up}} G_i^A \otimes G_i^B \tag{3.40}$$

- *Normalize the updated operator X_{up} and compute the new maximum operator Schmidt coefficient $\mu_{1,\text{up}}$*

Iterating those steps until the algorithm converges, will lead to a witness with a smaller white noise robustness. Further, it is to mention that the last step is only necessary since, in practice, the normalization could be lost due to numerical uncertainties.

3.2.2   Optimization with Respect to The Schmidt Operators

The OSD witnesses found by the optimization in the previous subsection are not necessarily optimal yet, as Algorithm 3.1 only changes the OS coefficients but leaves the Schmidt operators invariant. Thus, we need another algorithm, optimizing with respect to the operators. First, consider again the OSD witness:

$$\mathcal{W}_{\text{OSD}} = \mu_1 \mathbb{1} - \sum_{i=1}^{d^2} \mu_i G_i^A \otimes G_i^B, \qquad (3.41)$$

with $d = \min(d_A, d_B)$. Without loss of generality, the following computation is done for G_i^A. In order to transform the matrices G_i^A such that they still form an ONB afterwards, it is natural to perform rotations. As the optimization updates should be small, this can be written as:

$$G_i^A \mapsto \sum_{k=1}^{d_A^2} \delta O_{ik}^A \tilde{G}_k^A, \qquad (3.42)$$

where δO^A describes an infinitesimal rotation matrix and is given by

$$\delta O^A = \mathbb{1} + \sum_{l=1}^{n_A} \epsilon_A^{(l)} g^{(l)}. \qquad (3.43)$$

The vector $\vec{\epsilon}_A = \varepsilon \vec{v}_A$ denotes the rotation direction, with $\varepsilon \ll 1$, and the $g^{(l)}$ are the generators of the SO(N) group, whose elements are orthogonal $N \times N$ matrices with determinant equal to one. The generators $g^{(l)}$ are a set of matrices needed in order to express any of those elements and therefore are antisymmetric $N \times N$ matrices with only two nonzero entries one and minus one. The number of generators of a group SO(N) is given by $\frac{N(N-1)}{2}$. For $N = 3$, for example, there are $\frac{3(3-1)}{2} = 3$ generators, given as follows:

$$g^{(1)} = \begin{pmatrix} 0 & 1 & 0 \\ -1 & 0 & 0 \\ 0 & 0 & 0 \end{pmatrix}, \quad g^{(2)} = \begin{pmatrix} 0 & 0 & 1 \\ 0 & 0 & 0 \\ -1 & 0 & 0 \end{pmatrix}, \quad g^{(3)} = \begin{pmatrix} 0 & 0 & 0 \\ 0 & 0 & -1 \\ 0 & 1 & 0 \end{pmatrix}. \tag{3.44}$$

Here, we have $N = d_A^2$ as for the Transformation (3.42) d_A^2-dimensional matrices are needed. Hence, there are $n_A = \frac{d_A^2(d_A^2-1)}{2}$ generators in this case. Note that if $d_A < d_B$ and therefore $d_A = d$, the operators $\{\tilde{G}_k^A\}$ in Transformation (3.42) correspond to the Schmidt operators $\{G_i^A\}$. If $d_A > d_B$ and thus $d_A > d$, one has to perform a basis completion of the set $\{G_i^A\}$ in order to obtain an ONB $\{\tilde{G}_k^A\}$.

Now, we insert Equation (3.43) into (3.42) to obtain an equation for the transformation of the matrices G_i^A:

$$G_i^A \mapsto \sum_{k=1}^{d_A^2} \left((\mathbb{1}_{ik} + \sum_{l=1}^{n_A} \epsilon_A^{(l)} g_{ik}^{(l)}) \tilde{G}_k^A \right), \tag{3.45}$$

$$G_i^A \mapsto \sum_{k=1}^{d_A^2} \mathbb{1}_{ik} \tilde{G}_k^A + \sum_{k=1}^{d_A^2} \left(\sum_{l=1}^{n_A} \epsilon_A^{(l)} g_{ik}^{(l)} \tilde{G}_k^A \right), \tag{3.46}$$

$$G_i^A \mapsto \tilde{G}_i^A + \sum_{k=1}^{d_A^2} \left(\sum_{l=1}^{n_A} \epsilon_A^{(l)} g_{ik}^{(l)} \tilde{G}_k^A \right). \tag{3.47}$$

Using that for $i \leq d^2$ the operators G_i^A and $\tilde{G}_i^A$ are equal and the transformation is only done for d^2 operators, we can replace $\tilde{G}_i^A = G_i^A$ in the last line. Further, writing the infinitesimal rotation direction as $\vec{\epsilon}_A = \varepsilon \vec{v}_A$, we obtain:

$$G_i^A \mapsto G_i^A + \varepsilon \sum_{k=1}^{d_A^2} \left(\sum_{l=1}^{n_A} v_A^{(l)} g_{ik}^{(l)} \tilde{G}_k^A \right), \tag{3.48}$$

$$G_i^A \mapsto G_i^A + \varepsilon \sum_{l=1}^{n_A} \left(\sum_{k=1}^{d_A^2} v_A^{(l)} g_{ik}^{(l)} \tilde{G}_k^A \right), \tag{3.49}$$

$$G_i^A \mapsto G_i^A + \varepsilon \sum_{l=1}^{n_A} \left(v_A^{(l)} \sum_{k=1}^{d_A^2} g_{ik}^{(l)} \tilde{G}_k^A \right). \tag{3.50}$$

Then, we can express the transformation as:

$$G_i^A \mapsto G_i^A + \varepsilon \sum_{l=1}^{n_A} v_A^{(l)} (\xi^A)_i^{(l)} \text{ ,with } (\xi^A)_i^{(l)} = \sum_{k=1}^{d_A^2} g_{ik}^{(l)} \tilde{G}_k^A. \tag{3.51}$$

This looks similar to the transformation of the OS coefficients in the previous sub-section and therefore it is natural to apply a gradient descent here as well. Still, this problem is not completely analogous to the optimization of the OS coefficients, as in this case the update is performed on matrices instead of coefficients. Thus, we need to interpret it in a different way.

Instead of taking p_{crit} as function of the matrices G_i^A that should be updated, one can interpret p_{crit} as a function of the vector $\vec{v}_A$ (or $\vec{\epsilon}_A = \varepsilon \vec{v}_A$), which gives the direction of the matrices' rotation. In other words: While in Subsection 3.2.1 the goal was to find the optimal OS coefficients such that $p_{\text{crit}}(\vec{\mu}_{\text{opt}})$ is minimal, now we aim to find an optimal rotation direction $\vec{\epsilon}_{A,\text{ opt}} = \varepsilon \vec{v}_{A,\text{ opt}}$ such that the rotated operators $\{G_i^A\}$ minimize the function $p_{\text{crit}}(G_1^A, G_2^A, ... G_{d^2}^A) = p_{\text{crit}}(\vec{\epsilon}_A)$.

Having caught the problem, the first task is to write down the function $p_{\text{crit}}(\vec{\epsilon}_A)$ that should be minimized. As a reminder, we first consider the initial expression of the white noise robustness:

$$p_{\text{crit}} = \frac{\text{Tr}(\mathcal{W}_{\text{OSD}}\sigma_{\text{sep}})}{\text{Tr}(\mathcal{W}_{\text{OSD}}\sigma_{\text{sep}}) - \text{Tr}(\mathcal{W}_{\text{OSD}}\rho_{\text{ent}})} \tag{3.52}$$

$$= \frac{\text{Tr}((\mu_1 \mathbb{1} - X)\sigma_{\text{sep}})}{\text{Tr}((\mu_1 \mathbb{1} - X)\sigma_{\text{sep}}) - \text{Tr}((\mu_1 \mathbb{1} - X)\rho_{\text{ent}})} \tag{3.53}$$

$$= \frac{\mu_1 - \text{Tr}(X\sigma_{\text{sep}})}{\text{Tr}(X\rho_{\text{ent}}) - \text{Tr}(X\sigma_{\text{sep}})}. \tag{3.54}$$

Since the optimization should be performed with respect to the Schmidt operators, all operators are expressed in their OSD:

$$X = \sum_i \mu_i G_i^A \otimes G_i^B, \tag{3.55}$$

$$\sigma_{\text{sep}} = \sum_i s_i H_i^A \otimes H_i^B, \tag{3.56}$$

$$\rho_{\text{ent}} = \sum_i r_i K_i^A \otimes K_i^B. \tag{3.57}$$

Then, we find:

$$p_{\text{crit}} = \frac{\mu_1 - \text{Tr}\left((\sum_i \mu_i G_i^A \otimes G_i^B)(\sum_j s_j H_j^A \otimes H_j^B)\right)}{\text{Tr}\left((\sum_i \mu_i G_i^A \otimes G_i^B)(\sum_j r_j K_j^A \otimes K_j^B)\right) - \text{Tr}\left((\sum_i \mu_i G_i^A \otimes G_i^B)(\sum_j s_j H_j^A \otimes H_j^B)\right)} \tag{3.58}$$

$$= \frac{\mu_1 - \sum_{ij} \mu_i s_j \text{Tr}(G_i^A H_j^A)\text{Tr}(G_i^B H_j^B)}{\sum_{ij} \mu_i r_j \text{Tr}(G_i^A K_j^A)\text{Tr}(G_i^B K_j^B) - \sum_{ij} \mu_i s_j \text{Tr}(G_i^A H_j^A)\text{Tr}(G_i^B H_j^B)}, \tag{3.59}$$

where we used the linearity of the trace, as well as $\text{Tr}(A \otimes B) = \text{Tr}(A)\text{Tr}(B)$. In order to keep the computation clear, we summarize some constant expressions to:

$$S_{ij}^B = \mu_i s_j \text{Tr}(G_i^B H_j^B) \tag{3.60}$$

$$R_{ij}^B = \mu_i r_j \text{Tr}(G_i^B K_j^B). \tag{3.61}$$

The noise robustness then reads

$$p_{\text{crit}} = \frac{\mu_1 - \sum_{ij} S_{ij}^B \text{Tr}(G_i^A H_j^A)}{\sum_{ij} R_{ij}^B \text{Tr}(G_i^A K_j^A) - \sum_{ij} S_{ij}^B \text{Tr}(G_i^A H_j^A)}. \tag{3.62}$$

This is still independent from the rotation direction $\vec{e}_A$, so we write

$$G_i^A \mapsto G_i^A + \sum_{l=1}^{n_A} \epsilon_A^{(l)} (\xi^A)_i^{(l)}, \tag{3.63}$$

which is the Transformation (3.51) where $\vec{e}_A = \varepsilon \vec{v}_A$. This yields

$$p_{\text{crit}} = \frac{\mu_1 - \tilde{S} - \sum_{ij} S_{ij}^B \text{Tr}\left(\sum_l \epsilon_A^{(l)} (\xi^A)_i^{(l)} H_j^A\right)}{\tilde{R} + \sum_{ij} R_{ij}^B \text{Tr}\left(\sum_l \epsilon_A^{(l)} (\xi^A)_i^{(l)} K_j^A\right) - \tilde{S} - \sum_{ij} S_{ij}^B \text{Tr}\left(\sum_l \epsilon_A^{(l)} (\xi^A)_i^{(l)} H_j^A\right)}, \tag{3.64}$$

where $\tilde{S}$ and $\tilde{R}$ describe the following summarized constant terms:

$$\tilde{S} = \sum_{ij} S_{ij}^B \text{Tr}(G_i^A H_j^A) = \sum_{ij} \mu_i s_j \text{Tr}(G_i^B H_j^B)\text{Tr}(G_i^A H_j^A) \tag{3.65}$$

$$\tilde{R} = \sum_{ij} R_{ij}^B \text{Tr}(G_i^A K_j^A) = \sum_{ij} \mu_i r_j \text{Tr}(G_i^B K_j^B)\text{Tr}(G_i^A K_j^A). \tag{3.66}$$

Equation (3.64) defines then the function $p_{\text{crit}}(\vec{\epsilon}_A)$, which should be minimized with respect to $\vec{\epsilon}_A$. Note that the transformation in Equation (3.63) does not change the function $p_{\text{crit}}(\vec{\epsilon}_A)$ as it simply corresponds to the expression in Equation (3.62) with $\vec{\epsilon}_A = \vec{0}$.

So, we can compute the gradient using the quotient rule for derivatives:

$$p_{\text{crit}} = \frac{f}{g} \tag{3.67}$$

$$\Rightarrow p'_{\text{crit}} = \frac{g \cdot f' - g' \cdot f}{g^2}. \tag{3.68}$$

First, consider the numerator f:

$$f = \mu_1 - \tilde{S} - \sum_{ij} S_{ij}^B \text{Tr}\left(\sum_l \epsilon_A^{(l)}(\xi^A)_i^{(l)} H_j^A\right) \tag{3.69}$$

$$= \mu_1 - \tilde{S} - \sum_{ijl} S_{ij}^B \epsilon_A^{(l)} \text{Tr}\left((\xi^A)_i^{(l)} H_j^A\right). \tag{3.70}$$

The derivative of f with respect to $\epsilon_A^{(m)}$ is

$$f' = \frac{df}{d\epsilon_A^{(m)}} = -\sum_{ij} S_{ij}^B \text{Tr}\left((\xi^A)_i^{(m)} H_j^A\right). \tag{3.71}$$

The denominator can be treated analogously:

$$g = \tilde{R} + \sum_{ij} R_{ij}^B \text{Tr}\left(\sum_l \epsilon_A^{(l)}(\xi^A)_i^{(l)} K_j^A\right) - \tilde{S} - \sum_{ij} S_{ij}^B \text{Tr}\left(\sum_l \epsilon_A^{(l)}(\xi^A)_i^{(l)} H_j^A\right) \tag{3.72}$$

$$= \tilde{R} + \sum_{ijl} R_{ij}^B \epsilon_A^{(l)} \text{Tr}\left((\xi^A)_i^{(l)} K_j^A\right) - \tilde{S} - \sum_{ijl} S_{ij}^B \epsilon_A^{(l)} \text{Tr}\left((\xi^A)_i^{(l)} H_j^A\right) \tag{3.73}$$

and the derivative is:

$$g' = \frac{dg}{d\epsilon_A^{(m)}} = \sum_{ij} R_{ij}^B \text{Tr}\left((\xi^A)_i^{(m)} K_j^A\right) - \sum_{ij} S_{ij}^B \text{Tr}\left((\xi^A)_i^{(m)} H_j^A\right). \tag{3.74}$$

Now, having computed all terms appearing in Equation (3.68), the gradient can be found:

$$(\vec{\nabla}_{\tilde{\epsilon}}\, p_{\text{crit}}(\vec{\epsilon}))_m = \tag{3.75}$$

$$= \frac{\left(\tilde{R} + \sum_l \epsilon^{(l)} \tilde{R}_{\xi A}^{(l)} - \tilde{S} - \sum_l \epsilon^{(l)} \tilde{S}_{\xi A}^{(l)}\right) \cdot \left(-\tilde{S}_{\xi A}^{(m)}\right) - \left(\tilde{R}_{\xi A}^{(m)} - \tilde{S}_{\xi A}^{(m)}\right) \cdot \left(\mu_1 - \tilde{S} - \sum_l \epsilon^{(l)} \tilde{R}_{\xi A}^{(l)}\right)}{\left(\tilde{R} + \sum_l \epsilon^{(l)} \tilde{R}_{\xi A}^{(l)} - \tilde{S} - \sum_l \epsilon^{(l)} \tilde{S}_{\xi A}^{(l)}\right)^2},$$

$$\tag{3.76}$$

where we used the short notation $\vec{\epsilon}_A = \vec{\epsilon}$ and further defined:

$$\tilde{S}_{\xi A}^{(l)} = \sum_{ij} S_{ij}^B \text{Tr}\left((\xi^A)_i^{(l)} H_j^A\right) \tag{3.77}$$

$$\tilde{R}_{\xi A}^{(l)} = \sum_{ij} R_{ij}^B \text{Tr}\left((\xi^A)_i^{(l)} K_j^A\right). \tag{3.78}$$

However, as the original function p_{crit} is defined with the rotation direction equal to zero, we set $\vec{\epsilon}_A = \vec{0}$ again and obtain

$$(\vec{\nabla}_{\vec{\epsilon}_A}\, p_{\text{crit}}(\vec{\epsilon}_A))_m = \frac{\left(\tilde{R} - \tilde{S}\right) \cdot \left(-\tilde{S}_{\xi A}^{(m)}\right) - \left(\tilde{R}_{\xi A}^{(m)} - \tilde{S}_{\xi A}^{(m)}\right) \cdot \left(\mu_1 - \tilde{S}\right)}{\left(\tilde{R} - \tilde{S}\right)^2}. \tag{3.79}$$

The gradient $\vec{\nabla}_{\vec{\epsilon}_A}\, p_{\text{crit}}(\vec{\epsilon}_A)$ is then an $n_A = \frac{d_A^2(d_A^2-1)}{2}$-dimensional vector whose entries are given by Equation (3.79).

Next, we take a look at what actually happens to the operators G_i^A when performing the gradient descent algorithm. Starting with Transformation (3.63):

$$G_i^A \mapsto G_i^A + \sum_{l=1}^{n_A} \epsilon_A^{(l)} (\xi^A)_i^{(l)}, \tag{3.80}$$

the update performed in the gradient descent algorithm

$$\vec{\epsilon}_A \mapsto \vec{\epsilon}_A - \varepsilon \vec{\nabla}_{\vec{\epsilon}_A}\, p_{\text{crit}}(\vec{\epsilon}_A) \tag{3.81}$$

can be inserted:

$$G_i^A \mapsto G_i^A + \sum_{l=1}^{n_A} \left((\epsilon_A^{(l)} - \varepsilon(\vec{\nabla}_{\vec{\epsilon}_A} p_{\text{crit}}(\vec{\epsilon}_A))^{(l)})(\xi^A)_i^{(l)} \right) \tag{3.82}$$

$$G_i^A \mapsto G_i^A + \sum_{l=1}^{n_A} \epsilon_A^{(l)}(\xi^A)_i^{(l)} - \varepsilon \sum_{l=1}^{n_A} (\vec{\nabla}_{\vec{\epsilon}_A} p_{\text{crit}}(\vec{\epsilon}_A))^{(l)}(\xi^A)_i^{(l)}. \tag{3.83}$$

Thus, setting $\vec{\epsilon}_A = \vec{0}$ again, the final update rule for the operators G_i^A is found:

$$G_i^A \mapsto G_i^A - \varepsilon \sum_{l=1}^{n_A} (\vec{\nabla}_{\vec{\epsilon}_A} p_{\text{crit}}(\vec{\epsilon}_A))^{(l)}(\xi^A)_i^{(l)}. \tag{3.84}$$

Taking a closer look at it, one can note that the update rule corresponds to Transformation (3.51) with the negative gradient as rotation direction $\vec{v}_A = -\vec{\nabla}_{\vec{\epsilon}_A} p_{\text{crit}}(\vec{\epsilon}_A)$. Further, as mentioned before, the upper computations can be done analogously for the operators G_i^B receiving the update rule

$$G_i^B \mapsto G_i^B - \varepsilon \sum_{l=1}^{n_B} (\vec{\nabla}_{\vec{\epsilon}_B} p_{\text{crit}}(\vec{\epsilon}_B))^{(l)}(\xi^B)_i^{(l)}. \tag{3.85}$$

Finally, we can summarize the optimization with respect to the operators in a new algorithm:

Algorithm 3.2 (SO optimization for two parties).

- *Start with some input operator X_{in} and compute the OSD witness*
- *Write $X_{\text{in}} = \sum_i \mu_i G_i^A \otimes G_i^B$ as well as σ_{sep} and ρ_{ent} in their operator Schmidt decompositions*
- *Compute the gradient $\vec{\nabla}_{\epsilon_{A/B}} p_{\text{crit}}(\vec{\epsilon}_{A/B})$ of the white noise robustness ($\sigma_{\text{sep}} = \frac{1}{d^2}$)*
- *Update the Schmidt operators according to the update rule*

$$G_i^{A/B} \mapsto G_i^{A/B} - \varepsilon \sum_{l=1}^{n_{A/B}} (v_{A/B})^{(l)}(\xi^{A/B})_i^{(l)} \text{ with } \vec{v} = \frac{\vec{\nabla}_{\vec{\epsilon}_{A/B}} p_{\text{crit}}(\vec{\epsilon}_{A/B})}{\|\vec{\nabla}_{\vec{\epsilon}_{A/B}} p_{\text{crit}}(\vec{\epsilon}_{A/B})\|} \tag{3.86}$$

- Update the operator X_{in} using the new Schmidt operators:

$$X_{\text{up}} = \sum_i \mu_i G_{i,\text{up}}^A \otimes G_i^B \qquad (3.87)$$

or

$$X_{\text{up}} = \sum_i \mu_i G_i^A \otimes G_{i,\text{up}}^B \qquad (3.88)$$

- Normalize the updated operator X_{up} and compute the new maximum operator Schmidt coefficient $\mu_{1,\text{up}}$

Repeating those steps will give a minimum of the white noise robustness and thus, find an (improved) witness.

In order to obtain good results, it is necessary to make the updates not only on Alice's part (G_i^A) but also on Bob's part (G_i^B). A natural way to do so is changing between those two in each iteration. Thus, in the following, one iteration of Algorithm 3.2 is defined as the combination of one iteration for Alice's and one for Bob's part. Moreover, there is to mention that the last step of the algorithm is in theory not necessary, since the Schmidt operators should not change for one thing, and for the other the operator X_{up} should be normalized anyways as the update was done with a normalized gradient. However, in practice, there are numerical issues that make the last step necessary.

3.2.3 Practical Use of The Algorithms

In the last two subsections two algorithms to find OSD witnesses were presented. Now we want to take a look at these algorithms and check how to get optimal results in practice. For the following discussion the *unextendible product basis state* ρ_{UPB} is considered, which is a PPT entangled state and therefore can not be detected by fidelity witnesses. In order to define the state, the following five product vectors are needed [55]:

$$|\psi_0\rangle = \frac{1}{\sqrt{2}}|0\rangle(|0\rangle - |1\rangle) \tag{3.89}$$

$$|\psi_1\rangle = \frac{1}{\sqrt{2}}(|0\rangle - |1\rangle)|2\rangle \tag{3.90}$$

$$|\psi_2\rangle = \frac{1}{\sqrt{2}}|2\rangle(|1\rangle - |2\rangle) \tag{3.91}$$

$$|\psi_3\rangle = \frac{1}{\sqrt{2}}(|1\rangle - |2\rangle)|0\rangle \tag{3.92}$$

$$|\psi_4\rangle = \frac{1}{3}(|0\rangle + |1\rangle + |2\rangle)(|0\rangle + |1\rangle + |2\rangle). \tag{3.93}$$

They form a basis and since they are all pairwise orthogonal and there is no further product vector orthogonal to them, this basis is called *unextendible product basis (UPB)*. With this, the UPB state ρ_{UPB} can be defined:

$$\rho_{\text{UPB}} = \frac{1}{4}\left(\mathbb{1} - \sum_{i=0}^{4}|\psi_i\rangle\langle\psi_i|\right). \tag{3.94}$$

It is entangled due to the *range criterion* [23], which states that if a state σ is separable, its range is spanned by a set of product vectors $\{|a_i b_i\rangle\}$ and further, a set of product vectors $\{|a_i^* b_i\rangle\}$ spans the range of the partially transposed state σ^{T_A} [19].

Having introduced the UPB state, first, Algorithm 3.1 will be discussed. After that, we take a look at Algorithm 3.2 and lastly, several strategies to combine both algorithms are considered.

Algorithm 3.1 in Practice

Algorithm 3.1, which was introduced in the first part of this section, considers only the OS coefficients. Thus, applying it to find a witness for the UPB state is not expected to give the optimal[1] result unless one chooses a starting operator X_{in} with the optimal Schmidt operators. Since, in principle, the initial operator can be chosen arbitrarily, in the following, two options to choose X_{in} are considered. A natural choice is $X_{\text{in}}^{(1)} = \rho_{\text{UPB}}$ and thus the target state itself. This ansatz is compared to the choice $X_{\text{in}}^{(2)} = X_{\text{random}}$, where X_{random} is a random hermitian matrix.

---

[1] By optimal result we mean the smallest value p, such that the state $\tilde{\rho} = p\rho_{\text{UPB}} + (1-p)\frac{1}{9}$ is still detected.

So, the algorithm can be started using the following input parameters:

$$X_{\text{in}}^{(1)} = (1 - \varepsilon)\rho_{\text{UPB}} + \varepsilon\rho_{\text{random}} \text{ or } X_{\text{in}}^{(2)} = X_{\text{random}} \qquad (3.95)$$

$$\rho_{\text{ent}} = (1 - \varepsilon)\rho_{\text{UPB}} + \varepsilon\rho_{\text{random}} \qquad (3.96)$$

$$\sigma_{\text{sep}} = (1 - \varepsilon)\frac{1}{9} + \varepsilon\rho_{\text{random}}, \qquad (3.97)$$

with $\varepsilon = 0.0001$. The state ρ_{random} is a random density matrix and is added to all inputs, to make sure that the computer program performs the operator Schmidt decomposition within the complex numbers.[2] Now, Algorithm 3.1 is executed several times and after each iteration, the white noise robustness p_{crit} is computed. For the input operator $X_{\text{in}}^{(1)}$ the function of the white noise robustness converges after roughly 6.000 iterations. For the second ansatz $X_{\text{in}}^{(2)} = X_{\text{random}}$, the algorithm converges after about 60.000 iterations. Figure 3.1 shows the progress of the white noise robustness for both options $X_{\text{in}}^{(1)}$ and $X_{\text{in}}^{(2)}$ and one can see that the minimum values found with this algorithm are $p_{\text{crit}}^{(1)} = 0.8906$ for $X_{\text{in}}^{(1)}$ and $p_{\text{crit}}^{(2)} = 2.4054$ for $X_{\text{in}}^{(2)}$. Thus, an entanglement witness, detecting the UPB state, can be found using Algorithm 3.1, when taking the target state itself as initial input operator. However, there is to mention that, depending on the random matrix $X_{\text{in}}^{(2)}$ one chooses, it may be possible to find a witness that detects the state ρ_{UPB} with this ansatz as well. Further, also the number of iterations needed until the algorithm converges, strongly depends on the input operator.

Taking a closer look at the OS coefficients (see Figure 3.2) one can see that all of them become equal after a certain amount of iterations. It is to note that for the initial input operator $X_{\text{in}}^{(1)}$ the three smallest OS coefficients do not become equal to the others, as they were zero from the beginning. However, this behaviour is not surprising, as the OSD witness is a special form of the CCNR witness where all OS coefficients are equal to one. Therefore, the results for the white noise robustness p_{crit} can be compared to the smallest p, such that $\tilde{\rho}_{\text{UPB}} = p\rho_{\text{UPB}} + (1 - p)\frac{1}{9}$ is still detected by the CCNR criterion, which is $p^{\text{CCNR}} = 0.8822$ [14]. It is to mention that this is not the optimal value. The current bound is $p^{\text{opt}} = 0.87$ [43]. Further, using an optimization method called *Gilbert's algorithm*, in [54] it was shown that

---

[2] If $\varepsilon = 0$, computer programs automatically perform the OSD in the real numbers if the input state ρ is symmetric. This would not matter for the OS coefficients, but the Schmidt operators would not be hermitian in this case. Therefore, in order to break the symmetry and enforce an OSD in the complex numbers, some random noise is added to the inputs. This also ensures that the OSD is unique.

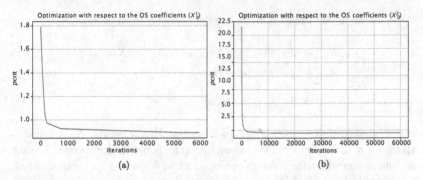

Figure 3.1 Progress of the white noise robustness using Algorithm 3.1 for the UPB state. Figure (a) shows the progress for the input operator $X_{\text{in}}^{(1)} \propto \rho_{\text{UPB}}$. The first value is $p_{\text{crit}} = 1.7846$ and thus, the state is not detected. After 6.000 iterations, a value $p_{\text{crit}} = 0.8906$ is obtained. So, a witness detecting ρ_{UPB} can be found. Figure (b) shows the progress for the input operator $X_{\text{in}}^{(2)} = X_{\text{random}}$. Here, the algorithm also converges, but ends at the value $p_{\text{crit}} = 2.4054$ and so, the witness does not detect ρ_{UPB}. Further, more iterations are needed until the algorithm converges

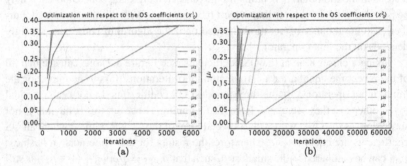

Figure 3.2 Progress of the OS coefficients using Algorithm 3.1 for the UPB state. Figure (a) shows the results for input operator $X_{\text{in}}^{(1)} \propto \rho_{\text{UPB}}$ and Figure (b) for input operator $X_{\text{in}}^{(2)} = X_{\text{random}}$. After a few iterations all OS coefficients are equal, besides those that were zero from the beginning

for $p^{\text{sep}} \leq 0.863$ the state ρ_{UPB} is separable. Considering again the result obtained by choosing $X_{\text{in}}^{(1)} \propto \rho_{\text{UPB}}$:

$$p_{\text{crit}}^{(1)} = 0.8906, \tag{3.98}$$

this corresponds to the value p^{CCNR} within the order of 10^{-3}. Noting that the input state ρ_{ent} was not the state ρ_{UPB} itself but mixed with noise at the order of 10^{-4}, one can say that $p_{\text{crit}}^{(1)} \approx p^{\text{CCNR}}$. Thus, a witness with the same white noise robustness as the optimal value for the CCNR criterion can be found, only optimizing the OS coefficients by starting with the initial operator $X_{\text{in}}^{(1)}$ equal to the target state. This, however, also is an expected result, as one can show that the CCNR witness is stronger than the OSD witness and the optimal choice of a witness that aims to detect a state $\rho_{\text{ent}} = \sum_i \mu_i \tilde{G}_i^A \otimes \tilde{G}_i^B$ is [3]:

$$\mathcal{W}_{\text{best}} = \mathbb{1} - \sum_i \tilde{G}_i^A \otimes \tilde{G}_i^B \tag{3.99}$$

The witness, found by Algorithm 3.1, is the following:

$$\mathcal{W}_{\text{UPB}}^{(1)} = \mu\mathbb{1} - \sum_i \mu\tilde{G}_i^A \otimes \tilde{G}_i^B = \mu\mathcal{W}_{\text{best}} \tag{3.100}$$

and thus, they are equal up to normalization. Therefore, the optimal OSD witness can be found using only Algorithm 3.1 by choosing a suitable input operator X_{in}. In the next subsection, we will consider Algorithm 3.2.

Algorithm 3.2 in Practice

The second algorithm optimizes the witness with respect to the Schmidt operators. Now we will check if taking the output operator $X_{\text{out}}^{(2)}$ from the first part, which was found by applying Algorithm 3.1 to a completely random input operator, as input of Algorithm 3.2, will lead to the optimal witness. Further, we consider again $X_{\text{in}}^{(1)} = \rho_{\text{UPB}}$ as input operator. As this algorithm only changes the Schmidt operators, we expect that the input operator $X_{\text{out}}^{(2)}$ will lead to the optimal witness, since its OS coefficients are already adapted. For ρ_{UPB} as input operator we expect that the optimal witness is not found because the OS coefficients differ and can not be influenced by the algorithm.

Indeed, applying Algorithm 3.2 to the input $X_{\text{out}}^{(2)}$ gives a witness with white noise robustness $p_{\text{crit}}^{(2)} = 0.8908 \approx p^{\text{CCNR}}$. This value is reached after 30.000 iterations. Taking $X_{\text{in}}^{(1)} \propto \rho_{\text{UPB}}$ as input, the algorithm converges after already 15.000 iterations resulting at the white noise robustness $p_{\text{crit}}^{(1)} = 1.7715$.

Thus, as expected, the optimal value is reached if the input operator's OS coefficients are all equal. If they are not equal, the optimal value is not found but less iterations are needed for convergence. This probably is the fact, since in principle, in this case the Schmidt operators are already optimal.

Summary and Strategies

Having tested the two algorithms for several input operators X_{in}, it was found that one only reaches the optimal witness if the input operator already has the correct Schmidt operators for Algorithm 3.1 or the optimal OS coefficients for Algorithm 3.2. So, there are two options to find the optimal witness for a completely random input operator:

1. Perform one algorithm until it converges and take the output operator as input for the second algorithm.
2. Combine Algorithm 3.1 and Algorithm 3.2 such that in each iteration first the Schmidt operators are updated and afterwards the OS coefficients or the other way around.

For the first strategy, we find that it makes a big difference which algorithm is applied first. For one thing, starting with the optimization of the Schmidt operators is more effortful, since this algorithm takes more time and iterations to converge. For the other thing, applying the SO optimization first, does not lead to the optimal witness in our example. This is the fact, since the algorithm searches the optimal Schmidt operators for given OS coefficients. However, if one starts with a completely random input operator X_{in}, the OS coefficients are not the optimal ones and thus the Schmidt operators corresponding to the optimal witness from Equation (3.99) can not be found.

One may wonder why applying Algorithm 3.1 to a completely random operator X_{in} still yields to the correct OS coefficients. This is due to the fact that choosing all OS coefficients equal is the optimal choice, independent of the Schmidt operators.

Applying the second option for the same random input operator, the algorithm converges after roughly 30.000 iterations and reaches the value $p_{\text{crit}}^{(2)} = 0.8904$. It is to mention that one iteration here corresponds to one iteration of the OS coefficient optimization and one iteration of the SO optimization. For the input operator ρ_{UPB} the minimum is reached after about 6.000 iterations.

In order to evaluate which strategy is more efficient, the number of iterations needed until the algorithm converges for both input states is summarized in Table 3.1. Further, in Table 3.2 the time needed for 1.000 iterations in each algorithm is denoted. Regarding Table 3.2, it is clear that the greatest effort comes from the SO optimization where the OSC optimization needs almost no time. It has to be mentioned that most of the time needed in Algorithm 3.1 is the time to plot the results which explains why the times for Algorithm 3.2 and the combination of the two algorithms do not differ a lot. Therefore, concerning Table 3.1, strategy 2 is the most efficient for a completely random input state. However, if the input state already has the correct Schmidt operators, strategy one is the best option since only the OS coefficients have to be found.

Table 3.1 Comparison of the two strategies to combine Algorithm 3.1 and Algorithm 3.2. The number of iterations needed until the algorithm converges is given. The first summand is the number of iterations needed in Algorithm 3.1, the second for Algorithm 4.1

	Input operator	Iterations
Strategy 1	$X_{\text{in}}^{(1)} \propto \rho_{\text{UPB}}$	$6.000 + 0$
	$X_{\text{in}}^{(2)} = X_{\text{random}}$	$60.000 + 30.000$
Strategy 2	$X_{\text{in}}^{(1)} \propto \rho_{\text{UPB}}$	30.000
	$X_{\text{in}}^{(2)} = X_{\text{random}}$	6.000

Table 3.2 Computing times needed to perform 1.000 iterations for a two-qutrit state compared for the different algorithms

Algorithm 3.1	1.35 s
Algorithm 3.2	123.0 s
Combination of A 3.1 and A 3.2	123.8 s

Lastly, there is to note that it is already known how the optimal OSD witness looks for the bipartite case and therefore one may argue that in principle the algorithms are not needed. However, in the next chapter we will see that these algorithms can be used to improve entanglement witnesses for the multipartite case and thus it is important that it was already verified that they work as expected.

3.3   Summary

In this chapter we introduced the OSD witness for bipartite states, which is given by

$$\mathcal{W}_{\mathrm{OSD}} = \mu_1 \mathbb{1} - X, \tag{3.101}$$

where μ_1 is the largest operator Schmidt coefficient of the operator X. We proved that this is indeed a witness as it has a positive expectation value for all separable states. Moreover, we showed that it is strictly stronger than fidelity witnesses since it detects not only NPT entangled states but also states violating the CCNR criterion. Lastly, two algorithms, finding the best OSD witness for some target state ρ_{ent} using a random input operator X_{in}, were introduced. One algorithm optimizes with respect to the OS coefficients and the other one with respect to the Schmidt operators. It is known that the best OSD witness corresponds to the optimal CCNR witness with the white noise robustness $p^{\mathrm{CCNR}} = 0.8822$ and we saw that one strategy to find this witness for any input operator X_{in} is applying the OSC optimization first and after that performing the SO optimization to the obtained output operator. The second strategy that leads to this optimal witness is performing both algorithms alternately. Thus, two reliable optimization strategies were found, which can be generalized and used to improve multiparticle entanglement witnesses in the next chapter.

The OSD Witness for the Multipartite Case 4

In the previous chapter we showed that one can construct entanglement witnesses based on the Schmidt decomposition of operators. It was found that this type of witness is strictly stronger than fidelity witnesses but always weaker than (or equal to) CCNR witnesses. Interestingly, this construction method can be generalized easily to the multipartite case. So in this chapter, first, we will discuss how to construct the OSD witness for the multipartite case and after that generalize the optimization algorithms from the previous chapter in order to improve given fidelity witnesses.

Within this chapter we improve the fidelity witnesses in many steps trying different strategies. However, the two main algorithms are the OSC optimization (Algorithm 4.1) and the SO optimization (Algorithm 4.4), which will be combined in the end. The final results corresponding to the best optimization strategy (Scheme (4.69)) can be found in Table 4.3. Moreover, Figure 4.6 in the last subsection of this chapter gives a graphical representation of the improvement.

4.1 Construction for Multiparticle States

For bipartite states the OSD witness is given by

$$W_{OSD} = \mu_1 \mathbb{1} - X, \tag{4.1}$$

Supplementary Information The online version contains supplementary material available at https://doi.org/10.1007/978-3-658-43203-4_4.

where μ_1 is greatest OS coefficient of the operator X. Therefore, for the multipartite case the OSD witness is constructed as follows:

Theorem 4.1 (Multipartite OSD witnesses). *Let X be a hermitian $d^N \times d^N$ matrix, where N is the number of particles with dimension d. Then*

$$\mathcal{W}_{\text{OSD}}^m = \lambda \mathbb{1} - X \tag{4.2}$$

is an entanglement witness for an N-qudit state. The coefficient λ is the greatest OS coefficient of the operator X maximized over all possible bipartitions:

$$\lambda = \max_{\text{all bipartitions}} \{\max_i (\mu_i^\alpha)\} \tag{4.3}$$

That Theorem 4.1 describes indeed a multiparticle entanglement witness, can be shown analogously to the proof for the bipartite witness. As it is almost analogous to the bipartite case, in the following only a sketch of the proof is provided.

Proof. Starting with the condition that the expectation value of $\mathcal{W}^m$ has to be positive on all biseparable states, it follows

$$\lambda \overset{!}{\geq} \text{tr}(X\rho_{\text{bisep}}) \; \forall \, \rho_{\text{bisep}} \in B. \tag{4.4}$$

To find the smallest coefficient λ that fulfills Condition (4.4), one has to maximize the expectation value of $\mathcal{W}^m$ with respect to all biseparable states:

$$\lambda \geq \max_{\rho_{\text{bisep}} \in B} \text{Tr}(X\rho_{\text{bisep}}). \tag{4.5}$$

Those states can be written as convex combinations of bipartitions:

$$\rho_{\text{bisep}} = \sum_\alpha \left(\sum_i p_i^\alpha \rho_i^{\alpha_A} \otimes \rho_i^{\alpha_B} \right), \tag{4.6}$$

where α denotes the possible bipartitions and α_A and α_B the parties corresponding to the first and second part. For example for a three particle state we have

$$\alpha \in \{A|BC, \; B|AC, \; C|AB\}, \tag{4.7}$$

where α_A and α_B for the first bipartition are:

$$\alpha = A|BC \tag{4.8}$$
$$\alpha_A = A \tag{4.9}$$
$$\alpha_B = BC \tag{4.10}$$

or the other way around. However, without loss of generality, here α_A (which we call *Alice's part* in the following) is always the smaller part.

To maximize the expectation value of X over all biseparable states, we can make use of the fact that it suffices to consider the extreme points of the set, which are given by the states $\rho_{\text{bisep}}^{\text{extr}} = \rho^{\alpha_A} \otimes \rho^{\alpha_B}$ for each bipartition α. Then the computation can be done analogously to the bipartite case, ending up with:

$$\max_{\rho_{\text{bisep}} \in B} \text{Tr}(X \rho_{\text{bisep}}) \leq \max_{\alpha} \mu_1^\alpha. \tag{4.11}$$

This ensures, that choosing

$$\lambda \geq \max_{\alpha} \mu_1^\alpha \tag{4.12}$$

will lead to an observable $\mathcal{W}^m$ that is positive on all biseparable states. $\qquad\square$

Having generalized the OSD witness to the multipartite case, the next step is adapting the optimization algorithms introduced for the bipartite witness. This will be discussed in detail in the next two sections. Note that since we will consider only the multipartite OSD witness in this chapter, we will simply write $\mathcal{W}^m = \mathcal{W}$ in the following.

4.2 OSC Optimization in the Multipartite Case

In this section the optimization of a given multipartite entanglement witness with respect to the OS coefficients will be discussed in detail. First, we will generalize Algorithm 3.1 for bipartite witnesses. Then some further approaches to obtain better results with this optimization strategy are introduced.

4.2.1 Generalization of Algorithm 3.1

Considering Algorithm 3.1, the first step after choosing an input operator X_{in}, is performing the operator Schmidt decomposition. For two parties it is clear how to do so because there is only one bipartititon. However, for more than two parties there are more bipartitions and therefore more possibilities to decompose the input operator X_{in}. Since the OSD witness is determined by the largest OS coefficient with respect to all bipartitions, it is natural to choose the bipartition where the OS coefficient is maximal. We will call this bipartition the *critical bipartition* and denote it by α' in the following.

The next step is the computation of the gradient. This can be done as for the bipartite case and thus, the gradient is given by

$$(\vec{\nabla}_{\vec{\mu}^{\alpha'}}\, p_{\text{crit}}(\vec{\mu}^{\alpha'}))_j = \frac{\left(\delta_{1j} - \tilde{\sigma}_j^{\alpha'}\right)\left(\sum_i \mu_i^{\alpha'}(\tilde{\rho}_i^{\alpha'} - \tilde{\sigma}_i^{\alpha'})\right) - \left(\tilde{\rho}_j^{\alpha'} - \tilde{\sigma}_j^{\alpha'}\right)\left(\mu_1^{\alpha'} - \sum_i \mu_i^{\alpha'}\tilde{\sigma}_i^{\alpha'}\right)}{\left(\sum_i \mu_i^{\alpha'}(\tilde{\rho}_i^{\alpha'} - \tilde{\sigma}_i^{\alpha'})\right)^2},$$

$$(4.13)$$

where the coefficients $\tilde{\rho}_i^{\alpha'}$ and $\tilde{\sigma}_i^{\alpha'}$ are defined analogously:

$$\tilde{\rho}_i^{\alpha'} := \text{Tr}\left((G_i^{\alpha'_A} \otimes G_i^{\alpha'_B})\rho_{\text{ent}}\right) \qquad (4.14)$$

$$\tilde{\sigma}_i^{\alpha'} := \text{Tr}\left((G_i^{\alpha'_A} \otimes G_i^{\alpha'_B})\sigma_{\text{sep}}\right). \qquad (4.15)$$

The following steps of Algorithm 3.1 can also be applied the same way. However, in order to be sure that the output has a positive expectation value on all biseparable states, in the last step we decompose the updated operator X_{up} again in the OSD and determine the critical bipartition of the new operator. Summarizing, this yields:

Algorithm 4.1 (OSC optimization for the multipartite case).

- *Start with some input operator X_{in}, find the critical bipartition α' and compute the OSD witness*
- *Write $X_{\text{in}} = \sum_i \mu_i^{\alpha'} G_i^{\alpha'_A} \otimes G_i^{\alpha'_B}$ in the operator Schmidt decomposition with respect to the critical bipartition*
- *Compute the gradient $\vec{\nabla}_{\vec{\mu}^{\alpha'}}\, p_{\text{crit}}(\vec{\mu}^{\alpha'})$ of the white noise robustness ($\sigma_{\text{sep}} = \frac{1}{d^N}$)*
- *Update the operator Schmidt coefficients according to the update rule*

$$\vec{\mu}^{\alpha'} \mapsto \vec{\mu}^{\alpha'} - \varepsilon \vec{v}^{\alpha'} \ \text{with} \ \vec{v}^{\alpha'} = \frac{\vec{\nabla}_{\vec{\mu}^{\alpha'}} \, p_{\text{crit}}(\vec{\mu}^{\alpha'})}{\|\vec{\nabla}_{\vec{\mu}^{\alpha'}} \, p_{\text{crit}}(\vec{\mu}^{\alpha'})\|} \tag{4.16}$$

- *Update the operator X_{in} using the new operator Schmidt coefficients:*

$$X_{\text{up}} = \sum_i \mu_{i,\text{up}}^{\alpha'} G_i^{\alpha'_A} \otimes G_i^{\alpha'_B} \tag{4.17}$$

- *Normalize the updated operator X_{up} and compute the new maximum operator Schmidt coefficient $\mu_{1,\text{up}}^{\alpha'_{\text{up}}}$. Take the new critical bipartition α'_{up} as input for the next iteration.*

Having adapted the algorithm to the multipartite case, we will discuss some examples now. First, we consider the four-qubit *Dicke state* with two excitations. Dicke states are entangled states, which were first investigated by R. H. Dicke in 1954 [11]. Symmetric Dicke states are of the form [19]:

$$|D_{k,N}\rangle = \binom{N}{k}^{-\frac{1}{2}} \sum_i P_i(|1\rangle^{\otimes k} \otimes |0\rangle^{\otimes N-k}), \tag{4.18}$$

with N denoting the number of particles and k the number of excitations. The notation $P_i(\cdot)$ describes a permutation of the qubits, where the sum runs over all possible permutations. Hence, for $k = 2$ excitations and $N = 4$ qubits we have:

$$|D_{2,4}\rangle = \frac{1}{\sqrt{6}}(|0011\rangle + |1100\rangle + |0110\rangle + |1001\rangle + |1010\rangle + |0101\rangle). \tag{4.19}$$

Dicke states with only $k = 1$ excitation are called *W states* [59].

As seen in the previous chapter, starting the optimization with the target state itself as input operator X_{in} leads to good results. Thus, we apply the algorithm, using the following input parameters:

$$X_{\text{in}} = (1 - \varepsilon)|D_{2,4}\rangle\langle D_{2,4}| + \varepsilon \rho_{\text{random}} \tag{4.20}$$

$$\rho_{\text{ent}} = (1 - \varepsilon)|D_{2,4}\rangle\langle D_{2,4}| + \varepsilon \rho_{\text{random}} \tag{4.21}$$

$$\sigma_{\text{sep}} = (1 - \varepsilon)\frac{1}{16} + \varepsilon \rho_{\text{random}}, \tag{4.22}$$

with $\varepsilon = 0.0001$ and ρ_{random} a random density matrix, which is added again in order to ensure a unique OSD. Clearly, the witness in step zero is equal to the fidelity witness:

$$\mathcal{W}_{\text{OSD},0} = \mu_1^{\alpha'} \mathbb{1} - X_{\text{in}} \tag{4.23}$$

$$= (s_1^{\alpha'})^2 \mathbb{1} - |D_{2,4}\rangle\langle D_{2,4}| \tag{4.24}$$

$$\Rightarrow \mathcal{W}_{\text{OSD},0} = \mathcal{W}_{\text{fid}}. \tag{4.25}$$

Therefore, Algorithm 4.1 can be used to improve the fidelity witness for a given state ρ_{ent}. Starting the optimization, the initial value of the white noise robustness is $p_{\text{crit}}^{\text{fid}} = 0.6445$. After roughly 2.500 iterations the algorithm converges, finding the minimum $p_{\text{crit}} = 0.5428$ (see Figure 4.1).

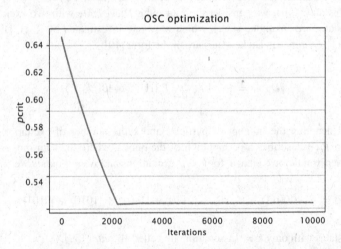

Figure 4.1 Progress of the white noise robustness using Algorithm 4.1 for the state $|D_{2,4}\rangle$. The first value, which corresponds to the white noise robustness of the fidelity witness, is $p_{\text{crit}} = 0.6445$. The algorithm converges straightforwardly to the value $p_{\text{crit}} = 0.5428$

However, for the four-qubit W state, the algorithm does not converge to a minimum. At the first 700 iterations the white noise robustness decreases and reaches a minimum of $p_{\text{crit}} = 0.7257$. After that, it increases and converges after about 22.000 iterations in total to the value $p_{\text{crit}} = 0.8137$. This value is worse than the white noise robustness for the fidelity witness, which is $p_{\text{crit}}^{\text{fid}} = 0.7334$. The progress of the white noise robustness is shown in Figure 4.2. This result seems strange, since usually, if the learning rate is chosen reasonably, a gradient decent algorithm should converge to a value smaller than the starting value and not increase. However, tracking the critical bipartition during the optimization, it is found that it changes almost with every iteration. This has not only the side effect that, different from the bipartite case, the Schmidt operators change as well but also explains why the white noise robustness does not decrease reliably: The gradient and therefore the direction of the steepest descent is computed with respect to the critical bipartition. However, after updating the OS coefficients, a new operator X_{up} is obtained, which has different OS coefficients in all bipartitions. Thus, the coefficient $\lambda = \mu_1^{\alpha'}$ does not change according to the update rule, which causes the fact that the white noise robustness increases in some iteration steps.

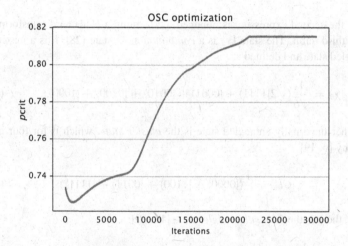

Figure 4.2 Progress of the white noise robustness using Algorithm 4.1 for the state $|W_4\rangle = |D_{1,4}\rangle$. After decreasing to the value $p_{\mathrm{crit}} = 0.7257$, the white noise robustness increases and converges to the value $p_{\mathrm{crit}} = 0.8137 > p_{\mathrm{crit}}^{\mathrm{fid}}$

Accordingly, the minimum found by the algorithm may be good for some bipartitions but not for every. However, the algorithm is constructed in such a way that it looks for the optimal value with respect to all possible bipartitions, which may be a value worse than for the fidelity witness. Nevertheless, this seems not to be a problem for all states, but only[1] for the W states. Later, in Subsection 4.2.3, we will

---

[1] With respect to the states we tested, the problem only occurred for the W states. However, there may be further states with this behavior.

investigate a strategy to better combine the optimization algorithms with respect to all bipartitions but for now, the goal is to improve the fidelity witness and therefore the convergence of the algorithm is of lower priority.

In the following, we applied the algorithm to several multipartite entangled states and the results are summarized in Table 4.1. In the table some states appear, which were not mentioned yet. The first one is a *hypergraph state* given by [29, 34]:

$$|Hyp\rangle = \frac{1}{\sqrt{8}}(|000\rangle + |001\rangle + |010\rangle + |011\rangle + |100\rangle + |101\rangle + |110\rangle - |111\rangle)$$

(4.26)

$$= \frac{1}{2}(|000\rangle + |010\rangle + |100\rangle + |111\rangle),$$ (4.27)

where the second expression is obtained, performing a Hadamard transformation of the third qubit. The state $|\chi\rangle$ is a *comb monotone* state [28]. It is a maximally entangled state and defined as

$$|\chi\rangle = \frac{1}{\sqrt{6}}(\sqrt{2}|1111\rangle + |0001\rangle + |0010\rangle + |0100\rangle + |1000\rangle).$$ (4.28)

A further maximally entangled state is the *cluster state*, which is for four qubits given by [6, 19]

$$|Cl_4\rangle = \frac{1}{2}(|0000\rangle + |1100\rangle + |0011\rangle - |1111\rangle).$$ (4.29)

Lastly, there is the *four-qubit singlet state* [19, 60]:

$$|\Psi_2\rangle = \frac{1}{\sqrt{3}}(|0011\rangle + |1100\rangle - \frac{1}{2}(|01\rangle + |10\rangle) \otimes (|01\rangle + |10\rangle)).$$ (4.30)

It is to mention that the GHZ state does not appear in Table 4.1 since it is known that the fidelity witness is optimal for the GHZ state [18].

Considering the results, one can see that for some states the white noise robustness has improved quite well. Especially for the hypergraph state, the four-qubit singlet state and the Dicke state $|D_{2,4}\rangle$, the white noise robustness gets at least 0.1 smaller. For the state $|\chi\rangle$ and the cluster state the white noise robustness did not change at all and for the W states, the value gets better but only at the order of 0.02. Further, the W states are the only ones, for which the algorithm does not converge to a minimum smaller than for the fidelity witness.

Table 4.1 Minimal white noise robustness found applying Algorithm 4.1, compared to the value for the fidelity witness for several states. For those states, which are marked with a (*), the algorithm does not converge to a minimum

State	$p_{\text{crit}}^{\text{fid}}$	$p_{\text{crit}}^{\text{OSD}}$
$\lvert Hyp \rangle$	0.714	0.558
$\lvert W_3 \rangle^*$	0.620	0.600
$\lvert \chi \rangle$	0.467	0.467
$\lvert Cl_4 \rangle$	0.467	0.467
$\lvert \Psi_2 \rangle$	0.733	0.572
$\lvert W_4 \rangle^*$	0.733	0.726
$\lvert D_{2,4} \rangle$	0.645	0.543
$\lvert D_{3,6} \rangle$	0.594	0.577

Although the algorithm improved the fidelity witness significantly for several states, the values are still far of the ones obtained considering PPT mixtures. Therefore, in the next subsection, some ideas to adapt the OSC optimization in order to obtain better results are introduced.

4.2.2   Further Approaches to Improve Fidelity Witnesses

As mentioned before, in this subsection we will discuss two ideas to further improve the results from the OSC optimization. The first idea is based on the fact that the direction of the gradient may not be the optimal direction to find a minimum. The second idea aims at the problem that the algorithm may get stuck in local minima.

Changing the Descent Direction

In order to minimize a function with respect to some parameter vector $\vec{\theta}$, one performs updates on this vector by taking infinitesimal steps in some direction $\vec{v}$. Since the gradient is the direction of the steepest slope, it is natural to go into the negative direction of the gradient, which is known as the gradient descent algorithm. However, although this is the direction of the steepest descent, it may not be the optimal direction. The algorithm could take far longer to find the minimum or even get stuck in a local one.

One simple idea to change the direction is adding some random orthogonal vector $\vec{v}_{\text{random}}^{\perp}$ to the gradient. Then, the new optimization direction is

$$\vec{v}_{\text{new}} = \vec{v} \pm \vec{v}_{\text{random}}^{\perp} = \frac{\vec{\nabla}_{\vec{\mu}}\, p_{\text{crit}}(\vec{\mu})}{\|\vec{\nabla}_{\vec{\mu}}\, p_{\text{crit}}(\vec{\mu})\|} \pm \vec{v}_{\text{random}}^{\perp}, \tag{4.31}$$

where we use the short hand notation $\vec{\mu} = \vec{\mu}^{\alpha'}$. To find a random vector $\vec{v}_{\text{random}}^{\perp}$, which is orthogonal to the gradient, first we generate a vector $\vec{v}_{\text{random}}$ whose entries are uniformly distributed. After that, the Gram-Schmidt process is performed with the gradient vector and the random vector as input. This ensures that the random vector is orthogonal to the direction of the gradient.

Summarizing, we can give the modified algorithm, where we noted the modifications by the * item:

Algorithm 4.2 (Modification of Algorithm 4.1 with changed descent direction).

- *Start with some input operator X_{in}, find the critical bipartition α' and compute the OSD witness*
- *Write $X_{\text{in}} = \sum_i \mu_i^{\alpha'} G_i^{\alpha'_A} \otimes G_i^{\alpha'_B}$ in the operator Schmidt decomposition with respect to the critical bipartition*
- *Compute the gradient $\vec{\nabla}_{\vec{\mu}^{\alpha'}}\, p_{\text{crit}}(\vec{\mu}^{\alpha'})$ of the white noise robustness ($\sigma_{\text{sep}} = \frac{1}{d^N}$)*
- *Generate a random vector $\vec{v}_{\text{random}}$ with uniformly distributed entries*
- *Perform the Gram-Schmidt process to the gradient and the vector $\vec{v}_{\text{random}}$ to obtain a vector orthogonal to the gradient $\vec{v}_{\text{random}}^{\perp}$*
- *Determine the new optimization direction by adding or subtracting the orthogonal vector to the normalized gradient $\vec{v}$:*

$$\vec{v}_{\text{new}} = \vec{v} \pm \vec{v}_{\text{random}}^{\perp} \tag{4.32}$$

- *Update the operator Schmidt coefficients according to the update rule*

$$\vec{\mu} \mapsto \vec{\mu} - \varepsilon \frac{\vec{v}_{\text{new}}}{\|\vec{v}_{\text{new}}\|} \tag{4.33}$$

- *Update the operator X_{in} using the new operator Schmidt coefficients:*

$$X_{\text{up}} = \sum_i \mu_{i,\text{up}}^{\alpha'} G_i^{\alpha'_A} \otimes G_i^{\alpha'_B} \tag{4.34}$$

- *Normalize the updated operator X_{up} and compute the new maximum operator Schmidt coefficient $\mu_{1,\text{up}}^{\alpha'_{\text{up}}}$. Take the new critical bipartition α'_{up} as input for the next iteration.*

The entries of the vector $\vec{v}_{\text{random}}$ are, as already mentioned, taken from a uniform distribution. However, the interval can be chosen arbitrarily and therefore, here we investigate the intervals $[-0.5, 0.5]$ and $[0, 1]$. In the following example we used the three-qubit W state, since here the white noise robustness has not improved much yet and the optimization is faster for less qubits. Applying the modified updates

$$\vec{\mu} \mapsto \vec{\mu} - \varepsilon \frac{\vec{v}_{\text{new}}}{\|\vec{v}_{\text{new}}\|}, \tag{4.35}$$

where $\vec{v}_{\text{new}}$ is defined by Equation (4.31) and $\vec{v}_{\text{random}}$ is another one in every update, leads to the following results:

$$p_{\text{crit, min}}^{[0,1]^-} = 0.5841 \tag{4.36}$$

$$p_{\text{crit, min}}^{[0,1]^+} = 0.5998 \tag{4.37}$$

$$p_{\text{crit, min}}^{[-0.5,0.5]^{\pm}} = 0.5998 \tag{4.38}$$

$$p_{\text{crit, min}} = 0.5996. \tag{4.39}$$

The indices $+$, $-$ and $\pm$ denote if the orthogonal random vector was added or subtracted from the gradient. In Figure 4.3 the progress of the white noise robustness is shown for the different random vector intervals. One can see that the progress of $p_{\text{crit}}^{[-0.5,0.5]^{\pm}}$ is similar to the unmodified algorithm, where $\vec{v}_{\text{random}}^{\perp} = \vec{0}$. It is to mention that it does not make a difference whether the orthogonal random vector is added or subtracted if the entries of the initial random vector are from a symmetric interval.

For $\vec{v}_{\text{random}}$ from the interval $[0, 1]^-$, the white noise robustness reaches a new minimum. So, this seems to be a direction, where the local minimum at $p_{\text{crit, min}} = 0.5996 \approx 0.600$ can be escaped and a new minimum at $p_{\text{crit, min}}^{[0,1]^-} = 0.5841 \approx 0.584$ is found.

This modification is now applied to all of the previous example states and we find that the white noise robustness improves for the four-qubit W state as well, leading to the minimum $p_{\text{crit, min}}^{[0,1]^-} = 0.722$. However, for all other states the minimal value either is the same as for the unmodified algorithm or worse. It should be noted that those results are only obtained when the initial random vector has entries

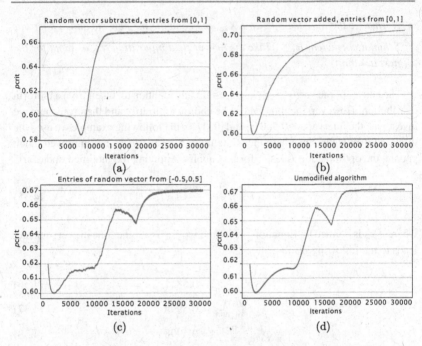

Figure 4.3 Progress of the white noise robustness for Algorithm 4.2 (Modification of Algorithm 4.1 with changed descent direction) applied to the three-qubit W state. Figures (a)–(c) show the progress for the different options of the random vector $\vec{v}_{\text{random}}$. Figure (d) shows the results for the unmodified algorithm

from the interval [0, 1] and the new direction vector is computed by subtracting the orthogonal random vector. Therefore, this direction seems to be a good one when improving the fidelity witnesses of W states.

Jumping Over Local Minima

Since the OSD witness is defined as

$$\mathcal{W}_{\text{OSD}} = \mu_1^{\alpha'} \mathbb{1} - X \tag{4.40}$$

with some arbitrary operator X consisting of a positive and a negative part, the idea came up to replace the operator X with its positive part X^+ in order to improve the witness [2]. Although it turned out that this approach only sometimes gives a better witness, the idea can be used to obtain better results in the optimization process.

First, in Algorithm 4.1 each witness, that is computed, is projected to the witness $\mathcal{W}_{\text{OSD}}^+$:

$$\mathcal{W}_{\text{OSD}} = \mu_1 \mathbb{1} - X \to \mathcal{W}_{\text{OSD}}^+ = \mu_1^+ \mathbb{1} - X^+, \qquad (4.41)$$

where we use again the short notation $\mu_1^{\alpha'} = \mu_1$. The operator X^+ is the positive part of X:

$$X = X^+ - X^- \qquad (4.42)$$

and X^- is the negative part. The coefficient μ_1^+ is the largest OS coefficient of X^+. Now, we apply Algorithm 4.1 to the three-qubit W state, but replace in each step:

$$X \to X^+ \qquad (4.43)$$

$$\mu_1 \to \mu_1^+. \qquad (4.44)$$

This causes the algorithm to find the minimum $p_{\text{crit, min}} = 0.587$, which is roughly the same as in the previous ansatz. Further, the white noise robustness also increases after reaching this minimum and converges to the value $p_{\text{crit}} = 0.603$ (see Figure 4.4).

The same is done for the four-qubit W state, where the minimum $p_{\text{crit, min}} = 0.724$ is found. For the other states, however, the projected positive witness $\mathcal{W}_{\text{OSD}}^+$ never gives a better white noise robustness.

Lastly, one may think that projecting to the positive witness $\mathcal{W}_{\text{OSD}}^+$ in every step might miss some good results since we have seen before that taking the positive part of X is not always the better choice. Therefore, we propose another adaption of Algorithm 4.1, noted again by the * items. In the following, Algorithm 4.1 is performed as usual, but in every step the white noise robustness p_{crit} is computed for both, the initial witness $\mathcal{W}_{\text{OSD}}$ and its positive projection $\mathcal{W}_{\text{OSD}}^+$. Then, whenever the white noise robustness is smaller for $\mathcal{W}_{\text{OSD}}^+$, we replace the operator X by its positive part X^+. This has the effect that the algorithm, when searching for the minimum, sometimes jumps to a different point of the function and therefore is able to find new local minima.

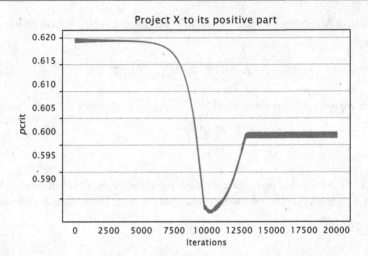

Figure 4.4 Progress of the white noise robustness for the modified Algorithm 4.1, where in each step the witness is projected to the positive part, applied to the three-qubit W state

Algorithm 4.3 (Modification of Algorithm 4.1 with implemented jumps).

- *Start with some input operator X_{in}, find the critical bipartition α' and compute the OSD witness*
- *Write $X_{\text{in}} = \sum_i \mu_i^{\alpha'} G_i^{\alpha'_A} \otimes G_i^{\alpha'_B}$ in the operator Schmidt decomposition with respect to the critical bipartition*
- *Compute the gradient $\vec{\nabla}_{\vec{\mu}^{\alpha'}} p_{\text{crit}}(\vec{\mu}^{\alpha'})$ of the white noise robustness ($\sigma_{\text{sep}} = \frac{1}{d^N}$)*
- *Update the operator Schmidt coefficients according to the update rule*

$$\vec{\mu}^{\alpha'} \mapsto \vec{\mu}^{\alpha'} - \varepsilon \vec{v}^{\alpha'} \quad \text{with} \quad \vec{v}^{\alpha'} = \frac{\vec{\nabla}_{\vec{\mu}^{\alpha'}} p_{\text{crit}}(\vec{\mu}^{\alpha'})}{\|\vec{\nabla}_{\vec{\mu}^{\alpha'}} p_{\text{crit}}(\vec{\mu}^{\alpha'})\|} \tag{4.45}$$

- *Update the operator X_{in} using the new operator Schmidt coefficients:*

$$X_{\text{up}} = \sum_i \mu_{i,\text{up}}^{\alpha'} G_i^{\alpha'_A} \otimes G_i^{\alpha'_B} \tag{4.46}$$

- *Normalize the updated operator X_{up} and compute the new maximum operator Schmidt coefficient $\mu_{1,\text{up}}^{\alpha'_{\text{up}}}$.*

* *Take the positive part of X_{up} and determine the witness* $\mathcal{W}_{\text{OSD}}^+ = (\mu_{1,\text{up}}^{\alpha'_{\text{up}}})^+ \mathbb{1} - X_{\text{up}}^+$.
* *Compute the white noise robustness* p_{crit} *and* p_{crit}^+ *for both witnesses.*
* *If* $p_{\text{crit}}^+ < p_{\text{crit}}$, *replace*

$$X_{\text{up}} \to X_{\text{up}}^+ \tag{4.47}$$

and take the new critical bipartition $(\alpha'_{\text{up}})^+$ *as input for the next iteration. Else continue with* X_{up} *and the corresponding coefficient* $\mu_{1,\text{up}}^{\alpha'_{\text{up}}}$.

Indeed, we find new minima for the W states:

$$p_{\text{crit, min}}^{|W_3\rangle} = 0.567 \tag{4.48}$$

$$p_{\text{crit, min}}^{|W_4\rangle} = 0.716. \tag{4.49}$$

For the three-qubit W state this value is 5.5 % better than the value found for the unmodified algorithm and for the four-qubit W state, the white noise robustness improves by 1.4 %. Figure 4.5 shows the progress of the algorithm compared to the projection to the positive witness $\mathcal{W}_{\text{OSD}}^+$.

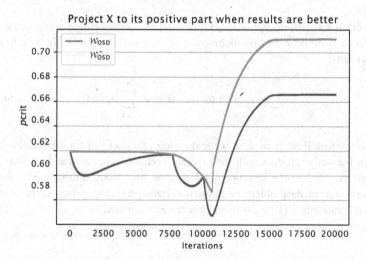

Figure 4.5 Progress of Algorithm 4.3 (Modification of Algorithm 4.1 with implemented jumps) for the three-qubit W state. The progress of the white noise robustness in the modified algorithm is compared to the white noise robustness of the projection to $\mathcal{W}_{\text{OSD}}^+$

One can see that the white noise robustness of $\mathcal{W}_{\mathrm{OSD}}^+$ is always greater than the white noise robustness of $\mathcal{W}_{\mathrm{OSD}}$. Whenever the noise robustness increases, the replacement $X \to X^+$ prevents the algorithm from finding the local minimum with a noise robustness worse than for the fidelity witness. However, at some point, both values p_{crit} and its projection p_{crit}^+ increase and converge to another local minimum greater than the values for the fidelity witness.

4.2.3 Multiple Gradient Descent Algorithm

As mentioned before, the goal is to optimize the OSD witness by updating the operator X. Since for the multipartite case there are multiple possibilities to decompose the operator into two parts, we look for an X that is optimal with respect to all bipartitions. In order to do so, another approach, based on the *multiple gradient descent algorithm (MGDA)* [10] is discussed now.

Consider the case where one has several criteria $J_i(Y)$ with $i = 1, ..., n$, which should be optimized concurrently. The $J_i(Y)$ are smooth functions of a vector $Y \in \mathbb{R}^N$. The idea of the MGDA is to generalize the gradient descent method to the multiobjective context by taking a direction that is a common descent direction for all criteria $J_i(Y)$ aiming to find a vector Y_0 where all criteria are *Pareto-stationary*. The term *Pareto-stationary* is defined as follows [10]:

Definition 4.1 (Pareto-stationarity). *The point Y_0 is called Pareto-stationary if and only if there exists a convex combination of the gradient vectors $\{\vec{u}_i = \vec{\nabla} J_i(Y_0)\}$ that is zero:*

$$\sum_{i=1}^{n} \alpha_i \vec{u}_i \text{ with } \alpha_i \geq 0 \ \forall i \text{ and } \sum_{i=1}^{n} \alpha_i = 1 \qquad (4.50)$$

In the gradient descent method we make use of the fact that for all vectors $\tilde{\vec{\omega}}$ for which the scalar product with the gradient vector is positive, one finds that $-\tilde{\vec{\omega}}$ is a direction where the loss function decreases. Analogous, for the multiobjective gradient descent, the problem is equivalent to finding a vector $\vec{\omega}$ whose scalar product with all gradients $J_i(Y)$ is positive. This can be expressed as

$$\vec{u}_i \cdot \vec{\omega} \geq 0 \ \forall i = 1, ..., n, \qquad (4.51)$$

where $\vec{u}_i = \frac{\vec{\nabla} J_i(Y_0)}{S_i}$ is the normalized gradient direction with some normalization factor S_i. Defining the convex hull of the family of vectors $\vec{u}_i$:

$$\overline{U} = \{\vec{u} \in \mathbb{R}^N | \vec{u} = \sum_{i=1}^{n} \alpha_i \vec{u}_i, \ \alpha_i \geq 0 \ \forall i = 1, ...n \text{ and } \sum_{i=1}^{n} \alpha_i = 1\}, \qquad (4.52)$$

one can show that the descent direction is given by

$$\vec{\omega} = \underset{\vec{u} \in \overline{U}}{\text{argmin}} \|\vec{u}\|, \qquad (4.53)$$

which is the element $\vec{u}$ of $\overline{U}$ with the smallest norm. The MGDA therefore works as follows [10]:

- Compute the gradient directions $\vec{u}_i$ and determine $\vec{\omega}$. If $\vec{\omega} \approx \vec{0}$, the point Y_0 is already Pareto-stationary and therefore the algorithm is done. Otherwise:
- Update the vector Y_0 according to

$$Y_0 \mapsto Y_0 - h\vec{\omega}, \qquad (4.54)$$

  where h is the step size (or learning rate). It is chosen as the greatest positive number, such that all functions $j_i(t) = J_i(Y_0 - t\vec{\omega})$ are monotone-decreasing in the interval $[0, h]$.

We may try to apply this algorithm to our optimization problem as the operator X (and therefore the OSD witness) should be optimal with respect to all bipartitions. Thus, the different bipartitions can be interpreted as the criteria $J_i(Y)$. However, it is not easy to define the $J_i(Y)$ since it should be different functions depending on the same vector Y. For example, taking simply $\vec{\nabla}_{\vec{\mu}^\alpha} p_{\text{crit}}(\vec{\mu}^\alpha)$ for all possible α as gradients $\vec{u}_i$ does not work, since in this case the vectors $\vec{\mu}^\alpha$ are different.

Also, if we found a way to describe the criteria $J_i(Y)$, we would still have to deal with the fact that the optimization is performed in operator space. One possibility to do so would be that, having found the gradient vectors $\vec{u}_i$, one would have to bring them back in operator form: $\vec{u}_i \rightarrow O_i$. So, having determined the O_i, the common descent direction $\vec{\omega} \rightarrow O$, which will also be in operator form then, can be computed. The scalar product, which we need in order to obtain the norm of O, becomes $\vec{\omega} \cdot \vec{\omega} \rightarrow \text{Tr}(O^\dagger O)$. Lastly, we would perform the update according to

$$X_{in} \mapsto X_{in} - hO. \tag{4.55}$$

However, we have not solved yet the task to define the criteria $J_i(Y)$ and therefore we can not test if the idea of embedding the vectors $\vec{u}_i$ into the operators O_i works out.

4.2.4   Short Summary

In this section the goal was to generalize the optimization algorithm with respect to the OS coefficients to the multipartite case. It turned out that the algorithm can be generalized easily by simply applying Algorithm 3.1 to the critical bipartition. However, we found that for some states, like the W states, the algorithm does not converge to a minimum. This could be explained by the fact that the entanglement witness needs to be optimized with respect to all bipartitions and therefore, changing the critical bipartition by making an update makes the value of the white noise robustness worse. Nevertheless, for almost all of the example states a witness with a white noise robustness smaller than for the fidelity witness was found.

However, since the improvement was quite low for some states, we tested two ideas of modifying the algorithm. One was to change the optimization direction by adding (or subtracting) a vector orthogonal to the gradient. The other idea was to project the entanglement witness $\mathcal{W}_{OSD}$ to the positive witness $\mathcal{W}_{OSD}^+$ whenever this would lead to a better white noise robustness, which causes the algorithm to jump to another position of the function and therefore find other minima. We saw that those two strategies only improved the witness for the W states, where the current minima for the three-qubit W state and the four-qubit W state are:

$$p_{crit,\,min}^{|W_3\rangle} = 0.567 \tag{4.56}$$

$$p_{crit,\,min}^{|W_4\rangle} = 0.716. \tag{4.57}$$

Both were reached by the last strategy of jumping during the optimization. For all other example states, the currently best results (see Table 4.1) are obtained using the unmodified Algorithm 4.1. However, these values are still not the optimal ones. This motivates, to also generalize the optimization algorithm with respect to the Schmidt operators (Algorithm 3.2), which will be done in the next section.

Lastly, there is to mention that another method to optimize multipartite witnesses for all possible bipartitions at once, the MGDA, was introduced. However, it turned out that it is not easy to apply the MGDA to our problem.

4.3   SO Optimization in the Multipartite Case

For the bipartite case the optimization with respect to the Schmidt operators was necessary since optimizing the OS coefficients did not change the operators at all. However, as mentioned before, in the multipartite case the OSC optimization changes the critical bipartition and therefore also the Schmidt operators do not stay unchanged. Nevertheless, this change is just a consequence of the OSC optimization and thus it is sensible to construct an algorithm really aiming to find the optimal Schmidt operators. So, in the following we will generalize the SO optimization to the multipartite case. After that, we will test it, using the previous example states and further discuss some strategies to combine the algorithms in order to obtain optimal results.

4.3.1   Generalization of Algorithm 3.2

The optimization with respect to the Schmidt operators can be generalized to the multipartite case analogously to the optimization with respect to the OS coefficients. In order to do so, we start again with determining the critical bipartition α' and decomposing the input operator X_{in} accordingly:

$$X_{\text{in}} = \sum_i \mu_i^{\alpha'} G_i^{\alpha'_A} \otimes G_i^{\alpha'_B}. \tag{4.58}$$

Different from the bipartite case, where only $d_A \times d_B$ systems with $d_A = d_B$ were considered, here, it may be possible that the operators $G_i^{\alpha'_A}$ and $G_i^{\alpha'_B}$ are of different dimensions. As mentioned before, without loss of generality, we choose Alice's system to be the smaller one: $d_{\alpha'_A} \leq d_{\alpha'_B}$. Consequently, if the critical bipartition is not symmetric and the rotation is done with respect to Bob's part:

$$G_i^{\alpha'_B} \mapsto \sum_{k=1}^{d_{\alpha'_B}^2} \delta O_{ik}^{\alpha'_B} \tilde{G}_k^{\alpha'_B}, \tag{4.59}$$

the $G_i^{\alpha'_B}$ do not form a basis, as there are only $d_{\alpha'_A}^2$ operators. Therefore, a basis completion has to be performed which is done by the Gram-Schmidt process. Then, the updates can be applied as in the bipartite case, just with respect to the critical bipartition:

$$G_i^{\alpha'_A} \mapsto G_i^{\alpha'_A} - \varepsilon \sum_{l=1}^{n_{\alpha'_A}} (\vec{\nabla}_{\vec{\epsilon}_{\alpha'_A}} \, p_{\text{crit}}(\vec{\epsilon}_{\alpha'_A}))^{(l)} (\xi^{\alpha'_A})_i^{(l)} \tag{4.60}$$

$$G_i^{\alpha'_B} \mapsto G_i^{\alpha'_B} - \varepsilon \sum_{l=1}^{n_{\alpha'_B}} (\vec{\nabla}_{\vec{\epsilon}_{\alpha'_B}} \, p_{\text{crit}}(\vec{\epsilon}_{\alpha'_B}))^{(l)} (\xi^{\alpha'_B})_i^{(l)}, \tag{4.61}$$

where the sum runs over $n_{\alpha'_{A/B}} = \dfrac{d_{\alpha'_{A/B}}^2 \left(d_{\alpha'_{A/B}}^2 - 1 \right)}{2}$ terms. The generalized algorithm reads then:

Algorithm 4.4 (SO optimization for the multipartite case).

- *Start with some input operator X_{in}, find the critical bipartition α' and compute the OSD witness*
- *Write $X_{\text{in}} = \sum_i \mu_i^{\alpha'} G_i^{\alpha'_A} \otimes G_i^{\alpha'_B}$ as well as σ_{sep} and ρ_{ent} in their operator Schmidt decompositions*
- *If $d_{\alpha'_A} < d_{\alpha'_B}$, perform a basis completion on the $G_i^{\alpha'_B}$*
- *Compute the gradient $\vec{\nabla}_{\vec{\epsilon}_{\alpha'_B}} \, p_{\text{crit}}(\vec{\epsilon}_{\alpha'_B})$ of the white noise robustness ($\sigma_{\text{sep}} = \frac{1}{d^N}$)*
- *Update the Schmidt operators according to the update rule*

$$G_i^{\alpha'_B} \mapsto G_i^{\alpha'_B} - \varepsilon \sum_{l=1}^{n_{\alpha'_B}} (v_{\alpha'_B})^{(l)} (\xi^{\alpha'_B})_i^{(l)} \text{ with } \vec{v} = \frac{\vec{\nabla}_{\vec{\epsilon}_{\alpha'_B}} \, p_{\text{crit}}(\vec{\epsilon}_{\alpha'_B})}{\|\vec{\nabla}_{\vec{\epsilon}_{\alpha'_B}} \, p_{\text{crit}}(\vec{\epsilon}_{\alpha'_B})\|} \tag{4.62}$$

- *Update the operator X_{in} using the new Schmidt operators:*

$$X_{\text{up}} = \sum_i \mu_i^{\alpha'} G_i^{\alpha'_A} \otimes G_{i,\text{up}}^{\alpha'_B} \tag{4.63}$$

- *Normalize the updated operator X_{up} and compute the new maximum operator Schmidt coefficient $\mu_{1,\text{up}}^{\alpha'_{\text{up}}}$. Take the new critical bipartition α'_{up} and repeat all steps for Alice's part. Here, a basis completion is not needed as Alice's part was defined to be the smaller one.*

It is to mention that one may also perform the updates only with respect to one part (Alice's or Bob's). However, this strategy does not work well. One reason is that optimizing with respect to Alice's part will almost change nothing since only a little part of the operator X_{in} is updated. Optimizing only with respect to Bob's part will also not lead to good results as only one part of the operator X_{in} will be improved.

In the following we apply the algorithm to the hypergraph state $|Hyp\rangle$ as well as the three-qubit W state $|W_3\rangle$. The input operator is again the projector of the states themselves mixed with some noise in order to break the symmetry and ensure a unique operator Schmidt decomposition. We find that for the hypergraph state the algorithm converges. However, although the updates were done with respect to both, Alice's and Bob's part, the white noise robustness only changed within the order of 10^{-4}. Furthermore, for the W state, the white noise robustness does not decrease at all but directly increases. It should be noted that it was not checked whether the algorithm converges at some point to a value greater than the white noise robustness for the fidelity witness. Nevertheless, one reason why the white noise robustness increases can be given. When updating the Schmidt operators with respect to one bipartition, the OS coefficients should stay unchanged, since the updated Schmidt operators still form orthonormal bases with respect to their systems Alice and Bob:

$$ X = \sum_i \mu_i^{\alpha'} G_i^{\alpha'_A} \otimes G_i^{\alpha'_B}, \tag{4.64} $$

$$ X_{\text{up}} = \sum_i \mu_i^{\alpha'} G_i^{\alpha'_A} \otimes G_{i,\,\text{up}}^{\alpha'_B} \tag{4.65} $$

or $\tag{4.66}$

$$ X_{\text{up}} = \sum_i \mu_i^{\alpha'} G_{i,\,\text{up}}^{\alpha'_A} \otimes G_i^{\alpha'_B}. \tag{4.67} $$

However, after the update, the OS coefficients of the other bipartitions may have changed and therefore also the critical bipartition could be another one. This leads to the same effect as in the OSC optimization: The gradient was computed with respect to the currently critical bipartition, after the update, however, the critical bipartition has changed and therefore the update direction may not have been the optimal one. In other words: Improving the witness with respect to one bipartition could make it worse for another one.

One idea to handle this problem, is to combine the algorithms and not only make updates with respect to the Schmidt operators but also with respect to the OS coefficients. In the next subsection strategies of combining the algorithms are investigated.

4.3.2   Combination Strategies

As seen before, the SO optimization (Algorithm 4.4) on its own does not lead to good results when trying to improve the fidelity witness. Therefore, we will investigate some strategies to combine it with the OSC optimization. As in the bipartite case, in principle there are two strategies. The first one is to start with one algorithm and iterate until it converges followed by applying the other one. The second strategy is to alternate between the optimization algorithms in each iteration.

For the first strategy is was already observed that starting the optimization with respect to the Schmidt operators does not give optimal results for the bipartite case. Further, as found in the previous part, starting with the SO optimization in the multipartite case, does not give good results, too. Therefore, the first strategy that will be checked, is beginning to apply the OSC optimization, and then optimizing with respect to the Schmidt operators. For those states, for which the white noise robustness converges to a minimum applying the OSC optimization, it is clear that we can take the output operator $X_{\text{out}}^{\text{OSC}}$ of the first algorithm as input for the operator optimization. However, for the W states the OSC optimization attains a minimum value smaller than the white noise robustness for the fidelity witness but then converges to a greater one. The naive way would be, to take that operator X^{min} as new input operator where the white noise robustness is minimal. Following this strategy and applying it to the W states $|W_3\rangle$ and $|W_4\rangle$, the hypergraph state $|Hyp\rangle$, the four-qubit Dicke state $|D_{2,4}\rangle$ and the four-qubit singlet state $|\Psi_2\rangle$, we find that the optimization of the Schmidt operators only improves the white noise robustness with respect to the fourth decimal place. Furthermore, the algorithm does not converge to a minimum but increases after a small improvement. It is to mention that we did not apply the SO optimization to the six-qubit states as the computation would be to effortful. In the Electronic Supplementary Material (Appendix B) we give a complete run-time analysis.

The second strategy, that should be investigated, is the idea of alternating between the algorithms. This means that in each iteration, the input operator X_{in} is updated with respect to Bob's Schmidt operators, then with respect to Alice's Schmidt operators and after that with respect to the OS coefficients. Starting with the initial operator $X_{\text{in}} \propto \rho_{\text{ent}}$, some results, better than those for only optimizing the OS coefficients, are found (see Table 4.2). Moreover, for all example states the algorithm converges to a minimum smaller than the value for the fidelity witness. Only for the three-qubit W state, the white noise robustness slightly increases after reaching the smallest value $p_{\text{crit, min}} = 0.581$ and then converges to the value $p_{\text{crit}} = 0.583$. Especially for the hypergraph state and the three-qubit W state the white noise robustness improved well compared to the values when only the OS coefficients were optimized. How-

Table 4.2 Results for applying Algorithm 4.1 and 4.4 alternately. The results for only optimizing the OS coefficients (Alg. 4.1) are compared to those where both algorithms are applied alternately. The values in the first column are those for the fidelity witness

State	$p_{\text{crit}}^{\text{fid}}$	$p_{\text{crit}}^{\text{OSD, OSC}}$	$p_{\text{crit}}^{\text{OSD, alt}}$
$|Hyp\rangle$	0.714	0.558	0.546
$|W_3\rangle$	0.620	0.600	0.581
$|\chi\rangle$	0.467	0.467	0.461
$|Cl_4\rangle$	0.467	0.467	0.463
$|\Psi_2\rangle$	0.733	0.572	0.572
$|W_4\rangle$	0.733	0.726	0.725
$|D_{2,4}\rangle$	0.645	0.543	0.540

ever, for other states there is almost no difference between the two optimization methods. Furthermore, it is to mention that the values for the W states are still worse than those, found in Subsection 4.2.2 by modifying the OSC optimization algorithm. Therefore, it is clear that this strategy only finds local minima.

Looking back to the bipartite case, the OSC optimization converged to an operator X where all OS coefficients are the same: $\mu_i = \mu \ \forall i$. Therefore, it is worth considering the Schmidt coefficients of the operators X in the multipartite case. Indeed, taking the output operator $X_{\text{out}}^{\text{OSC}}$ to which the algorithm converges for the three-qubit W state and the hypergraph state, we find that all OS coefficients are equal with respect to all bipartitions, meaning $\mu_i^\alpha = \mu \ \forall i \ \forall \alpha$. For the four-qubit states we find that the input operator X_{in} converges to an X where for the non-symmetrical bipartitions $(d_{\alpha_A} < d_{\alpha_B})$ all OS coefficients are equal. Further, the maximum OS coefficients for the other bipartitions are equal as well but may be different from the ones of the non-symmetrical ones. A reason for that could be that for four-partite states some bipartitions never become critical and consequently the algorithm converges when only the OS coefficients of the critical bipartitions are adjusted. It is to note that one can not simply perform the update with respect to a non-critical bipartition as the coefficient $\lambda = \mu_1^{\alpha'}$ appears in the formula for p_{crit} and therefore one would have to express $\mu_1^{\alpha'}$ in the μ_i^α of another bipartition.

Thus, the conjecture arises that the optimal operator X has to fulfill that all OS coefficients are equal with respect to all bipartitions. Based on this conjecture, it would be natural to apply the OSC optimization first, such that the OS coefficients fulfill the condition

$$\mu_i^\alpha = \mu \ \forall i \ \forall \alpha. \tag{4.68}$$

And then apply the SO optimization to this $X_{\text{out}}^{\text{OSC}}$ in order to find the correct Schmidt operators. However, this is exactly what was done previously for all states besides the W states and it was observed that there is no great improvement found, for one thing and that the white noise robustness does not decrease reliably, for the other thing. Yet, taking a closer look at what happens during the optimization, we find the following: As mentioned before, the SO optimization does not leave the Schmidt coefficients unchanged. Therefore, applying it to the operator $X_{\text{out}}^{\text{OSC}}$, where all OS coefficients are equal, will mix them up again and the witness gets worse. Consequently, the idea is to apply the alternating strategy to $X_{\text{out}}^{\text{OSC}}$, according to the scheme:

$$X_{\text{in}} \propto \rho_{\text{target}} \xrightarrow{\text{Alg. 3}} X_{\text{out}}^{\text{OSC}} \xrightarrow{\text{Alg. 3 \& Alg. 6, alt}} X_{\text{out}}^{\text{best}}. \tag{4.69}$$

This makes sure that after each update on the Schmidt operators the OS coefficients are adjusted such that they stay equal.

Applying this strategy, an algorithm is found, which converges to an improved white noise robustness for all example states besides the four-qubit singlet state $|\Psi_2\rangle$. The results are summarized in Table 4.3. One can see that the new minima found for the W states are better than the ones found with the modified OSC optimization algorithm. Additionally, the white noise robustness for the hypergraph state reached a new minimum, too. However, the four-qubit singlet state did not improve further. There is to mention that for the comb monotone state $|\chi\rangle$ and the four-qubit cluster state this strategy does not differ from the previous one since applying only the

Table 4.3 Best results, obtained by applying Alg. 4.1 and Alg. 4.4 alternately to the output operator of the OSC optimization (Alg. 4.1) according to Scheme (4.69). The smallest values found for the white noise robustness are compared to the values for the fidelity witness

State	$p_{\text{crit}}^{\text{fid}}$	$p_{\text{crit}}^{\text{OSD}}$
$|Hyp\rangle$	0.714	0.545
$|W_3\rangle$	0.620	0.556
$|\chi\rangle$	0.467	0.461
$|Cl_4\rangle$	0.467	0.463
$|\Psi_2\rangle$	0.733	0.572
$|W_4\rangle$	0.733	0.714
$|D_{2,4}\rangle$	0.645	0.540

OSC optimization did not change the white noise robustness nor the operator X_{in}. Also for the four-qubit Dicke state both strategies give the same result which might be caused by the fact that its OS coefficients for the symmetrical bipartitions are already equal. Therefore, for those states the values in Table 4.3 are equal to those in Table 4.2.

4.3.3 Short Summary

After generalizing the OSC optimization in the previous section, here the SO optimization was adapted to the multipartite case. We saw that the generalization can be done analogously by simply applying the algorithm for the bipartite case to the critical bipartition. However, while in the bipartite case Alice's and Bob's parts are of the same dimension and therefore the $G_i^{A/B}$ already form a basis, in the multipartite case it may happen that the critical bipartition is not symmetric and therefore one needs to complete the set of the G_i^B to a basis in order to perform the update on Bob's part. The generalized algorithm was then applied to the hypergraph state and the three-qubit W state, where the initial input operator X_{in} was the target state itself. Here we found that for the hypergraph state the white noise robustness only improves in the order of 10^{-4} and for the W state it does not improve at all but gets worse. Therefore, some strategies to combine this algorithm with the optimization of the OS coefficients were investigated.

Since we already knew that starting with the optimization of the Schmidt operators is not an effective strategy, the first idea was to start with the optimization of the OS coefficients and after that applying the SO optimization to that operator X^{min}, for which the white noise robustness is minimal. Following this strategy, we could improve the white noise robustness for some states but again only at the order of 10^{-4}. Furthermore, the algorithm did not converge to the minimum but increased after reaching it. Therefore, we applied another strategy, alternating between the optimization of Bob's operators, Alice's operators and the OS coefficients. The results obtained with this strategy can be found in Table 4.2. For almost all example states new minima were found. Only the value for the four-qubit singlet state could not be improved. However, the values for the W states were still larger than those, found by modifying the algorithm in Subsection 4.2.2, which confirms that the minima, found by this algorithm are only local ones.

Lastly, we applied another strategy, based on the idea that the OS coefficients of the optimal operator X might be equal with respect to all bipartitions (Condition (4.68)). Therefore, we first optimized the OS coefficients and after that took the output operator X_{out}^{OSC} to which this algorithm converges as input for the next optimization, where we alternated between both optimization methods (Scheme 4.69). This lead to the final results in Table 4.3.

Most of those results can be compared to (optimal) values obtained by using *PPT mixtures* [39]. PPT mixtures are states that can be written in the form:

$$\rho^{\mathrm{PPT\,mix}} = \sum_{\alpha} p^{\alpha} \rho^{\alpha}_{\mathrm{PPT}}, \tag{4.70}$$

where $\rho^{\alpha}_{\mathrm{PPT}}$ are PPT states that are separable with respect to the bipartition α. The set of these states is an approximation of the set of biseparable states and can be used to detect genuine multipartite entanglement. In [39], the white noise robustness $p^{\mathrm{PPT\,mix}}$ was determined for the W states, the four-qubit Dicke state, the four-qubit cluster state and the four-qubit singlet state. Furthermore, it was verified that the values for the three-qubit W state and the cluster state are optimal in the sense that adding more white noise to these states would lead to a biseparable one. In [29], $p^{\mathrm{PPT\,mix}}$ was computed for the hypergraph state and it was found that the obtained value is optimal, too.

Figure 4.6 compares these (optimal) values to our results and the white noise robustness for the fidelity witness. Note, that for the state $|\chi\rangle$ the value $p^{\mathrm{PPT\,mix}}$ was not computed yet, but in principle this can be done analogously to the computations in [39]. One can see that the resulting values of the white noise robustness do still not reach the results from [39] and [29]. Nevertheless, for most states they improved quite well with respect to the values for the fidelity witness. Moreover, it is to mention that it may be possible that the OSD witness can not reach those values at all.

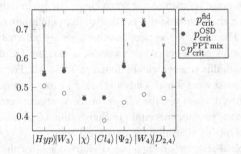

Figure 4.6 Final results for the white noise robustness, obtained using Scheme (4.69). The improvement of the fidelity witnesses is compared to the (partially optimal) results one obtains using PPT mixtures [39]

Schmidt Number Witnesses

5

The concept of the OSD witness can also be applied to certify the dimensionality of entanglement. In this chapter we will show how to construct Schmidt number witnesses using the scheme of the operator Schmidt decomposition. Furthermore, we will give an example.

5.1 Construction of Schmidt Number Witnesses

Using the concept introduced in Chapter 3, one can construct Schmidt number witnesses, which is shown in the following.

Theorem 5.1 (OSD Schmidt number witnesses). *Let X be a hermitian operator, then a k-Schmidt witness based on the OSD is given by*

$$\mathcal{W}_{kS} = \lambda_k \mathbb{1} - X. \tag{5.1}$$

The coefficient λ_k depends on the Schmidt number k and can be computed by

$$\lambda_k \geq \max_{|\Psi^{k-1}\rangle \in S_{k-1}} \langle \Psi^{k-1}|X|\Psi^{k-1}\rangle, \tag{5.2}$$

where $|\Psi^{k-1}\rangle$ denotes a pure state with Schmidt rank $k-1$.

Supplementary Information The online version contains supplementary material available at https://doi.org/10.1007/978-3-658-43203-4_5.

S. Denker, *Characterizing Multiparticle Entanglement Using the Schmidt Decomposition of Operators*, BestMasters,
https://doi.org/10.1007/978-3-658-43203-4_5

Equation (5.2) ensures that the expectation value of $\mathcal{W}_{kS}$ is positive on all states with Schmidt number $k - 1$. In the following, we will take a closer look at this equation and try to find the actual coefficients λ_k. For this, we start estimating the maximum analogously to the proof in Section 3.1.1:

$$\max_{|\Psi^{k-1}\rangle \in S_{k-1}} \langle \Psi^{k-1}|X|\Psi^{k-1}\rangle = \max_{|\Psi^{k-1}\rangle \in S_{k-1}} \mathrm{Tr}(X|\Psi^{k-1}\rangle\langle\Psi^{k-1}|). \tag{5.3}$$

The operators X and $|\Psi^{k-1}\rangle\langle\Psi^{k-1}|$ can be decomposed in their OSD and after analogous computations we end up with the following:

$$\max_{\tilde{s}_l} \mathrm{Tr}\left(\sum_j \mu_j G_j^A \otimes G_j^B \sum_l \tilde{s}_l H_l^A \otimes H_l^B\right) \leq \max_{\tilde{s}_l} \sum_{jl} \mu_j \tilde{s}_{l\downarrow}. \tag{5.4}$$

The $\tilde{s}_{l\downarrow}$ are the decreasingly ordered operator Schmidt coefficients of $|\Psi^{k-1}\rangle\langle\Psi^{k-1}|$ and therefore are given by the products of the vector Schmidt coefficients s_i of $|\Psi^{k-1}\rangle = \sum_{i=1}^{k-1} s_i|\alpha_i\beta_i\rangle$. In the following subsections we will consider Equation (5.4) for $k = 3$, $k = 4$ and $k > 4$.

5.1.1 Schmidt Number Witnesses for $k = 3$

We start with Schmidt number $k = 3$, which means that it is maximized over all states $|\Psi^2\rangle$ with Schmidt rank $k - 1 = 2$. Thus, the decreasingly ordered operator Schmidt coefficients $\tilde{s}_{l\downarrow}$ of the projector $|\Psi^2\rangle\langle\Psi^2|$ are given by the products of the vector Schmidt coefficients of the states $|\Psi^2\rangle$: $\{s_1 s_1, s_1 s_2, s_2 s_1, s_2 s_2\}$. The term in Equation (5.4) then simplifies to

$$\max_{\tilde{s}_l} \sum_{jl} \mu_j \tilde{s}_{l\downarrow} = \max_{s_1,s_2}(\mu_1 s_1^2 + \mu_2 s_1 s_2 + \mu_3 s_2 s_1 + \mu_4 s_2^2) = \max_{s_1,s_2} f(s_1, s_2),$$

$$\tag{5.5}$$

where the coefficients s_1 and s_2 have to fulfill $s_1^2 + s_2^2 = 1$, $s_1, s_2 > 0$ and $s_1 \geq s_2$. Taking a closer look at the function $f(s_1, s_2)$, one finds that it can also be written as a matrix vector multiplication

$$f(s_1, s_2) = \langle s|M|s\rangle \tag{5.6}$$

with the normalized vector $|s\rangle = (s_1, s_2)^T$ and the matrix

$$M = \begin{pmatrix} \mu_1 & \mu_2 \\ \mu_3 & \mu_4 \end{pmatrix}. \tag{5.7}$$

Since the matrix M is not necessarily symmetric, it needs to be symmetrized, such that Equation (5.4) describes an eigenvalue problem, which is solved by the largest eigenvalue of the matrix $M^{\text{symm}} = \frac{M+M^T}{2}$:

$$\max_{|s\rangle}\langle s|M|s\rangle = \max_{|s\rangle}\langle s|M^{\text{symm}}|s\rangle. \tag{5.8}$$

Then, the eigenvalues γ can be computed, which for a 2×2 matrix are given by:

$$\gamma_{1,2} = \frac{1}{2}\text{Tr}(M^{\text{symm}}) \pm \sqrt{\left(\frac{1}{2}\text{Tr}(M^{\text{symm}})\right)^2 - \det(M^{\text{symm}})} \tag{5.9}$$

$$\Rightarrow \gamma_{1,2} = \frac{\mu_1 + \mu_4}{2} \pm \sqrt{\left(\frac{\mu_1 + \mu_4}{2}\right)^2 - \mu_1\mu_4 + \left(\frac{\mu_2 + \mu_3}{2}\right)^2} \tag{5.10}$$

$$= \frac{1}{2}(\mu_1 + \mu_4) \pm \frac{1}{2}\sqrt{(\mu_1 + \mu_4)^2 - 4\mu_1\mu_4 + (\mu_2 + \mu_3)^2} \tag{5.11}$$

$$= \frac{1}{2}\left(\mu_1 + \mu_4 \pm \sqrt{\mu_{14}^2 + \mu_{23}^2}\right). \tag{5.12}$$

In equation line (5.12) we used the short hand notations $\mu_{14} := \mu_1 - \mu_4$ and $\mu_{23} := \mu_2 + \mu_3$. Since all μ_i as well as μ_{23} and μ_{14} are positive, the maximum eigenvalue is

$$\gamma_{\text{max}} = \frac{1}{2}\left(\mu_1 + \mu_4 + \sqrt{\mu_{14}^2 + \mu_{23}^2}\right). \tag{5.13}$$

It is to note that the coefficients of the eigenvector $|s\rangle$ corresponding to the maximum eigenvalue of M^{symm} are all positive, since having a negative coefficient would make the expression $\langle s|M^{\text{symm}}|s\rangle$ smaller. Moreover, since $|s\rangle$ describes an eigenvector, it can be normalized and therefore $s_1^2 + s_2^2 = 1$ is fulfilled. The last condition, the s_i need to fulfill, is that they should be ordered decreasingly ($s_1 \geq s_2$). This, however is guaranteed by the way the matrix M and therefore M^{symm} is constructed. Since the operator Schmidt coefficients μ_i are also in decreasing order and therefore $\mu_1 \geq \mu_4$, the first entry s_1 of the eigenvector maximizing Expression (5.8) must be greater or equal to the second one: $s_1 \geq s_2$.

Thus, we can construct the Schmidt witness for Schmidt number $k = 3$,

$$\mathcal{W}_{3S} = \lambda_3 \mathbb{1} - X, \tag{5.14}$$

using that the coefficient λ_3 is given by γ_{max} in Equation (5.13) and the μ_1, μ_2, μ_3 and μ_4 are the four largest operator Schmidt coefficients of the operator X.

5.1.2 Schmidt Number Witnesses for $k = 4$

Now we want to find the coefficient λ_4 such that the Schmidt witness has a positive expectation value on all states $|\Psi^3\rangle$ with Schmidt number $k - 1 = 3$. Proceeding analogously to the case $k = 3$, the operator Schmidt coefficients of the projector $|\Psi^3\rangle\langle\Psi^3|$ are given by $\{\tilde{s}_l\} = \{s_1 s_1, s_1 s_2, s_2 s_1, s_1 s_3, s_3 s_1, s_2 s_2, s_2 s_3, s_3 s_2, s_3 s_3\}$. To obtain the function $f(s_1, s_2, s_3) = \sum_{jl} \mu_j \tilde{s}_{l\downarrow}$ the operator Schmidt coefficients $\tilde{s}_l$ have to be sorted decreasingly. However, there is no unique order, since depending on how the s_i look like, it is either $s_1 s_3 > s_2 s_2$ or $s_1 s_3 < s_2 s_2$. They also might be equal ($s_1 s_3 = s_2 s_2$), but then the order would not matter anyway. This leads to two possible functions $f(s_1, s_2, s_3)$ and $g(s_1, s_2, s_3)$ that have to be considered:

$$f(s_1, s_2, s_3) = \mu_1 s_1^2 + \mu_2 s_1 s_2 + \mu_3 s_2 s_1 + \mu_4 \underline{s_1 s_3} + \mu_5 \underline{s_3 s_1} + \mu_6 s_2^2$$
$$+ \mu_7 s_2 s_3 + \mu_8 s_3 s_2 + \mu_9 s_3^2 \tag{5.15}$$

$$g(s_1, s_2, s_3) = \mu_1 s_1^2 + \mu_2 s_1 s_2 + \mu_3 s_2 s_1 + \mu_4 \underline{s_2^2} + \mu_5 \underline{s_1 s_3} + \mu_6 \underline{s_3 s_1}$$
$$+ \mu_7 s_2 s_3 + \mu_8 s_3 s_2 + \mu_9 s_3^2. \tag{5.16}$$

The coefficient λ_4 is then given by the maximum of those two functions:

$$\max_{s_1, s_2, s_3} \left(f(s_1, s_2, s_3), g(s_1, s_2, s_3) \right). \tag{5.17}$$

This can be found interpreting $f(s_1, s_2, s_3)$ and $g(s_1, s_2, s_3)$ as an eigenvalue problems like in the previous subsection. The corresponding matrices are given by the following:

$$M_1 = \begin{pmatrix} \mu_1 & \mu_2 & \boxed{\mu_4} \\ \mu_3 & \boxed{\mu_6} & \mu_7 \\ \boxed{\mu_5} & \mu_8 & \mu_9 \end{pmatrix} \tag{5.18}$$

$$M_2 = \begin{pmatrix} \mu_1 & \mu_2 & \boxed{\mu_5} \\ \mu_3 & \boxed{\mu_4} & \mu_7 \\ \boxed{\mu_6} & \mu_8 & \mu_9 \end{pmatrix}. \tag{5.19}$$

Now M_1 and M_2 have to be symmetrized such that maximizing $f(s_1, s_2, s_3)$ and $g(s_1, s_2, s_3)$ describes an eigenvalue problem. This can be done by taking $M_n^{\text{symm}} = \frac{M_n + M_n^T}{2}$, $(n = 1, 2)$. Thus, the coefficient λ_4 is given by

$$\lambda_4 = \max(\text{maxEig}(M_1^{\text{symm}}), \text{maxEig}(M_2^{\text{symm}})), \tag{5.20}$$

where $\text{maxEig}(M_n^{\text{symm}})$ with $n = 1, 2$ describes the maximum eigenvalue of the symmetrized matrices $M_n \mapsto M_n^{\text{symm}}$ from Equations (5.18) and (5.19). Since the M_n are 3×3 matrices, λ_4 is the solution of a third order polynomial in the μ_i $(i = 1, ..., 9)$, which are the first nine operator Schmidt coefficients of the operator X. Therefore, also computing λ_4 is numerically easy.

5.1.3   Schmidt Number Witnesses for $k > 4$

In this subsection we consider Schmidt number witnesses for Schmidt number $k > 4$. As already seen in the previous subsection, for $k > 3$ the order of the operator Schmidt coefficients $\tilde{s}_l$ is not unique anymore and thus, for $k = 5$ there are already lots of matrices to check. In the following, the possible matrix arrangements will be considered and a formula that bounds the number of arrangements will be given.

We start by reformulating the actual problem: In order to obtain the coefficient λ_k, one needs to maximize a function given by

$$f(s_1, ..., s_{k-1}) = \sum_{i=1}^{(k-1)^2} \mu_i \tilde{s}_{i\downarrow}. \tag{5.21}$$

However, the order of the operator Schmidt coefficients $\tilde{s}_l$ depends on the vector Schmidt coefficients s_i. The task is now to arrange the $\tilde{s}_l = s_j s_{j'}$ in decreasing order or rather determine the number of possibilities N to do so. Those arrangements will then lead to matrices M_n, $(n = 1, ..., N)$ of the operator Schmidt coefficients μ_i

whose greatest eigenvalues have to be determined in order to obtain λ_k as the greatest of all those N maximum eigenvalues.

Having recalled this, first, the matrix M, that needs to be filled, is considered:

$$M = \begin{pmatrix} \cdots\cdots \\ \cdots\cdots \\ \cdots\cdots \\ \vdots\ \vdots\ \vdots\ \ddots \end{pmatrix}. \tag{5.22}$$

M is of the dimension $(k-1) \times (k-1)$ because there are $(k-1)^2$ summands in Equation (5.21). Since the matrix will be symmetrized and also some coefficients $\tilde{s}_l = s_j s_{j'} = s_{j'} s_j$ appear twice, it is sufficient to only consider the diagonal and the upper or lower triangular matrix:

$$M = \begin{pmatrix} \cdots\cdots \\ \ \cdots\cdots \\ \ \ \cdots\cdots \\ \ \ \ \ddots \end{pmatrix} \tag{5.23}$$

The number of entries in that part is given by

$$N_{\text{entries}} = \sum_{i=1}^{k-1} i = \frac{(k-1)k}{2}, \tag{5.24}$$

as there are $k-1$ rows with $k-1$ entries each but in every row one entry less is considered. The right hand side of the equation is found by applying the Gaussian sum formula. So, without knowing anything further, the task would be to arrange N_{entries} coefficients, which would lead according to the urn model to $N_{\text{entries}}!$ possibilities. However, since we know that $\tilde{s}_l = s_j s_{j'}$ and the s_i are sorted decreasingly, we can presort the products $s_j s_{j'}$ in the following manner:

$$\begin{matrix} s_1 s_1 & s_1 s_2 & s_1 s_3 & s_1 s_4 & \cdots \\ & s_2 s_2 & s_2 s_3 & s_2 s_4 & \cdots \\ & & s_3 s_3 & s_3 s_4 & \cdots \\ & & & s_4 s_4 & \cdots \\ & & & & \ddots \end{matrix} \tag{5.25}$$

In this array, the values decrease from the top to the bottom and from the left to the right. However, the relation on all diagonals is unknown in general.

Taking a closer look at the arrangement (5.25), it is found that this can be interpreted as a *shifted standard Young tableau*, if the Schmidt coefficients s_i differ pairwise. The only differences are that, first, in a shifted standard young tableau the values of the entries increase from the top to the bottom and from the left to the right instead of decreasing and second that in a shifted standard Young tableau one would have natural numbers $\{1, ..., N_{\text{entries}}\}$ as entries. This, however does not matter for the following considerations.

For every (shifted) Young diagram the number of arrangements of the entries is given by the *Hook length formula* [4]:

$$f_\lambda = \frac{n!}{\prod_{s \in \lambda} h_\lambda(s)}, \tag{5.26}$$

which will be explained in more detail in the Electronic Supplementary Material (Appendix C). We can use this formula to compute the number of possible arrangements of the $\tilde{s}_l$. Interestingly, the Young diagram in (5.25) has even a special shape, namely a *shifted staircase shape*. For Young diagrams of that shape the number of shifted standard Young tableaux and therefore the number of possible arrangements of $\tilde{s}_l$ is given by [26]:

$$g_\lambda = n! \prod_{i=0}^{m-1} \frac{i!}{(2i+1)!}, \tag{5.27}$$

which follows from Schur's product formula for shifted shapes [37, 53]. The factor n describes the total number of boxes or entries the Young diagram has, which is in this case the number of coefficients $\tilde{s}_l$ that should be arranged and therefore is given by $n = N_{\text{entries}} = \frac{(k-1)k}{2}$ as found above. The variable m is the number of the rows in the tableau and therefore is given by the Schmidt number $k - 1$. With this, an estimate of the number of possible arrangements of the $\tilde{s}_l$ can be given, which only depends on the Schmidt number k:

$$N = \left(\frac{(k-1)k}{2} \right)! \prod_{i=0}^{k-2} \frac{i!}{(2i+1)!}. \tag{5.28}$$

Now, we check this formula by comparing it to the actual number of possibilities found numerically. Considering the results (Table 5.1), one can see for one thing

Table 5.1 Number of possible arrangements of operator Schmidt coefficients found by using Formula (5.28) compared with a Python simulation for Schmidt numbers $k = 3$ to $k = 6$

k	$N_{formula}$	$N_{numerically}$
3	1	1
4	2	2
5	12	10
6	286	114

that for Schmidt number $k = 6$ already more than hundred 5×5 matrices can be found, which means that the computational effort highly increases with the Schmidt number. For the other thing, we observe that the numerical simulation differs from the formula, which means that there are less degrees of freedom than assumed.

This is due to the fact that the Schmidt coefficients have certain properties (they are normalized, positive and sorted decreasingly). Thus, deciding the relation between one pair $\tilde{s}_l$ and $\tilde{s}_{l'}$, fixes the relation between another one. As an example, consider Schmidt number $k = 5$. According to the presorted scheme in (5.25), twelve orders can be found. However, it is found that the following two orders are not possible, although they fulfill Scheme (5.25):

$$s_2 s_2 > s_1 s_3 > s_1 s_4 > s_2 s_3 > s_3 s_3 \not> s_2 s_4, \qquad (5.29)$$

$$s_1 s_3 > s_2 s_2 > s_2 s_3 > s_1 s_4 > s_2 s_4 \not> s_3 s_3. \qquad (5.30)$$

One can see this, considering the ratios of the vector Schmidt coefficients s_i [20]:

$$s_2 s_2 > s_1 s_3 \text{ and } s_1 s_4 > s_2 s_3 \Rightarrow 1 \leq \frac{s_2 s_2}{s_1 s_3} \cdot \frac{s_1 s_4}{s_2 s_3} = \frac{s_2 s_4}{s_3 s_3} \Rightarrow s_3 s_3 \leq s_2 s_4, \quad (5.31)$$

$$s_1 s_3 > s_2 s_2 \text{ and } s_2 s_3 > s_1 s_4 \Rightarrow 1 \geq \frac{s_2 s_2}{s_1 s_3} \cdot \frac{s_1 s_4}{s_2 s_3} = \frac{s_2 s_4}{s_3 s_3} \Rightarrow s_3 s_3 \geq s_2 s_4, \quad (5.32)$$

where the computation in the first row (equation line (5.31)) refers to the first excluded order (equation line (5.29)) and the second row to the second one, respectively. Also, there is to note that if the Schmidt coefficients s_i do not differ pairwise, there are even less possibilities to arrange the operator Schmidt coefficients $\tilde{s}_l$.

5.2 Example

In this section we give an example for the k-Schmidt witness based on the OSD. For this, one may consider the mixed two-ququad state:

$$\rho_3 = \frac{1}{2}|\Psi_+^3\rangle\langle\Psi_+^3| + \frac{1}{4}(|23\rangle + |32\rangle)(\langle 23| + \langle 32|), \qquad (5.33)$$

with

$$|\Psi_+^3\rangle = \frac{1}{\sqrt{3}}(|00\rangle + |11\rangle + |22\rangle). \qquad (5.34)$$

This state has Schmidt number three, but fidelity-based Schmidt witnesses only detect Schmidt number two [1], which is equivalent to certifying entanglement. The fidelity-based entanglement witness is given by:

$$\mathcal{W}_2 = \frac{1}{3}\mathbb{1} - |\Psi_+^3\rangle\langle\Psi_+^3|. \qquad (5.35)$$

It detects the state ρ_3 with the white noise robustness $p_{\text{crit}}^{\text{fid}} = 0.620$. Taking the optimal OSD witness, as given in Chapter 3 (Eq. (3.99)), the white noise robustness $p_{\text{crit}}^{\text{OSD}} = 0.357$ is found.

Further, using the scheme of the OSD, one can construct a witness certifying Schmidt number three. As found in the previous section, it is given by

$$\mathcal{W}_{3S} = \lambda_3 \mathbb{1} - X, \qquad (5.36)$$

with $\lambda_3 = \frac{1}{2}(\mu_1 + \mu_4 + \sqrt{\mu_{14}^2 + \mu_{23}^2})$. In Chapter 3 we noted that the best choice for the operator X is found by taking the Schmidt operators from the target state ρ_3 and setting all OS coefficients to one. Although it is not clear yet if this is also the best choice for Schmidt number witnesses, it is a natural approach to chose the operator X the same way here. This yields:

$$\lambda_3 = \frac{1}{2}\left(1 + 1 + \sqrt{0^2 + 2^2}\right) = 2. \qquad (5.37)$$

Consequently, the OSD 3-Schmidt witness is given by

$$\mathcal{W}_{3S} = 2\mathbb{1} - \sum_i \tilde{G}_i^A \otimes \tilde{G}_i^B, \tag{5.38}$$

where the $\tilde{G}_i^{A/B}$ are the Schmidt operators of ρ_3. This witness certifies Schmidt number three with the white noise robustness $p_{crit}^{3SOSD} = 0.830$.

It is to mention that if the choice of $X = \sum_i \tilde{G}_i^A \otimes \tilde{G}_i^B$ was indeed also optimal for the k-Schmidt witnesses, the coefficient λ_k would reduce to $\lambda_k = k - 1$. This is due to the fact that there would be only one possibility to arrange the OS coefficients in a matrix M, namely a $(k-1) \times (k-1)$ matrix with all entries equal to one. These matrices have only one nonzero eigenvalue which is given by

$$\mathrm{maxEig}(M) = \mathrm{Tr}(M) = (k-1) \cdot 1. \tag{5.39}$$

Hence, the OSD k-Schmidt witness would read

$$\tilde{\mathcal{W}}_{kS} = (k-1)\mathbb{1} - \sum_i \tilde{G}_i^A \otimes \tilde{G}_i^B. \tag{5.40}$$

Note that this corresponds to an extension of the CCNR criterion for the Schmidt number [25]. Furthermore, it is to mention that the fidelity witness for Schmidt numbers (Eq. 2.78), up to normalization, is equal to the OSD Schmidt witness if one chooses $X = |\Psi^+\rangle\langle\Psi^+| = \sum_i G_i \otimes G_i$, where $|\Psi^+\rangle$ is the maximally entangled state for any dimension and therefore $\tilde{G}_i^A = \tilde{G}_i^B = G_i \, \forall \, i$. However, if $G_i^A \neq G_i^B$, one can not say that the witnesses are equal.

5.3   Short Summary

Lastly, we give a short summary of this chapter. We showed that the concept of the OSD witnesses can be used to construct Schmidt number witnesses. They are given by:

$$\mathcal{W}_{kS} = \lambda_k \mathbb{1} - X, \tag{5.41}$$

where the coefficient λ_k can be found maximizing the expectation value of X with respect to all states with Schmidt number $k-1$. The computation of this maximum is completely analogous to the proof for the bipartite entanglement witness. However, one ends up with a function of the OS coefficients of X and the vector Schmidt coefficients of a state $|\Psi^{k-1}\rangle$. This function can be maximized easily for $k = 3$ by interpreting it as an eigenvalue problem. Thus, the coefficient λ_3 is given by the largest eigenvalue of a matrix M containing the four largest OS coefficients of X and therefore is simply the solution of a second order polynomial. However, for $k \geq 4$ it gets more complicated since the function is not unique and therefore one has to compute the maximum eigenvalues of several matrices.

In the end, we saw in an example that the OSD Schmidt number witness can detect Schmidt number three for the mixed state ρ_3 where the fidelity witness fails. Furthermore, we found that the fidelity witness for Schmidt numbers is a special case of the OSD Schmidt witness.

Conclusion and Outlook

6

In this thesis a new type of entanglement witnesses, based on the Schmidt decomposition of operators was introduced and investigated. They are of the form:

$$W = \lambda \mathbb{1} - X. \tag{6.1}$$

In Chapter 3 we showed how to construct these witnesses in the bipartite case, namely by choosing $\lambda = \mu_1$ as the largest OS coefficient of the operator X. Furthermore, we clarified that the OSD witness detects strictly more states than the fidelity witness as it certifies entanglement for states violating the PPT criterion as well as those violating the CCNR criterion. Then, two algorithms to optimize given entanglement witnesses in the bipartite case with respect to the white noise robustness were proposed. The first one (Algorithm 3.1) optimizes the OSD witness by searching the optimal OS coefficients of the operator X. The second one (Algorithm 3.2) improves the witness performing infinitesimal rotations of the Schmidt operators. Both algorithms are based on the gradient descent method. Although in the bipartite case it is clear which choice of the operator X is the optimal one in order to obtain the best entanglement witness for a given target state ρ_{ent}, the two algorithms were tested for the unextendible product basis state ρ_{UPB}. We found that applying first Algorithm 3.1 to a completely random OSD witness will lead to an operator X where all OS coefficients are equal. Applying Algorithm 3.2 afterwards to this output finds the best choice of the OSD witness, where all OS coefficients are equal and the Schmidt operators are the same as for the target state ρ_{UPB}. The same result is also found if one applies them alternately. Thus, the optimization algorithms lead to the most optimal CCNR witness for the UPB state, which confirms that they work properly.

S. Denker, *Characterizing Multiparticle Entanglement Using the Schmidt Decomposition of Operators*, BestMasters,
https://doi.org/10.1007/978-3-658-43203-4_6

In the next chapter we generalized the OSD witness to the multipartite case. It turned out that this can be done by simply taking the coefficient λ as the largest OS coefficient of the operator X with respect to all possible bipatitions. Then the optimization algorithms were adjusted. We found that both algorithms can be adapted to the multipartite case by taking the critical bipartition of the operator X and then performing the algorithms as in the bipartite case. We applied the OSC optimization algorithm to a hypergraph state, a four- and a six-qubit Dicke state, a three- and a four-qubit W state, a comb monotone state $|\chi\rangle$, a four-qubit cluster state and the four-qubit singlet state $|\Psi_2\rangle$ and found that especially for the hypergraph, the singlet and the four-qubit Dicke state the white noise robustness improved quite well. However, for the W states the improvement was not that good and also the algorithm converged to a value higher than the white noise robustness for the fidelity witness. Hence, we tested two approaches to find new minima, where one strategy was changing the descent direction and the other one was jumping over local minima. Indeed, we found new minima for the W states, using these strategies. Moreover, we introduced the MGDA as alternative optimization algorithm, but it was found that applying it to this problem is not trivial. Lastly, since the SO optimization on its own did not perform very well, we investigated some strategies to combine it with the OSC optimization. It turned out that the best strategy is to first optimize with respect to the OS coefficients and then, when the white noise robustness converges, apply both algorithms alternately, such that the Schmidt operators are improved for one thing but the OS coefficients are adjusted for the other thing (Scheme (4.69)). The results, found for the example states, are summarized in Table 4.3. Especially for the three-qubit W state the witness could be improved well compared to the previous results, when only optimizing the OS coefficients. However, also for the other states we could further improve the white noise robustness. Only for the four-qubit singlet state applying the algorithms alternately did not improve the witness anymore.

However, the values found for the white noise robustness are still worse than those, found by using PPT mixtures [39], which is shown in Figure 4.6. This leads to two open questions. For one thing, it would be interesting to know which kind of states the OSD witness can detect. For example for bipartite states it is clear that those states which violate the CCNR criterion and the PPT criterion can be detected by the OSD witnesses. However, in the multipartite case, it is only known that the OSD witnesses detect all states that are detected by fidelity witnesses.

A further question would be how the optimal operator X for the OSD witness looks like. As already mentioned, in the bipartite case the optimal choice is taking all OS coefficients equal and choosing the Schmidt operators equal to those of the target state. For the multipartite case we conjecture that the optimal choice would be an operator X whose OS coefficients are all equal with respect to all possible bipartitions, since those operators are found by the OSC optimization algorithm for three-qubit states. For four-qubit states the algorithm only finds operators whose OS coefficients with respect to the non-symmetrical bipartitions are equal. Hence, one might assume that the OSC optimization does not work optimal yet. One approach to overcome this problem might be the MGDA, introduced in Subsection 4.2.3. If one could find a way to define the criteria $J_i(Y)$ properly for each bipartition, the algorithm would probably find another operator X.

In the last chapter we showed how to construct Schmidt number witnesses. We found that the coefficient λ_k is the solution of a $k - 1$ order polynomial in the first k^2 OS coefficients of the operator X. These polynomials are found by computing the largest eigenvalue of a matrix of the OS coefficients. However, the arrangement of the entries in the matrix is not unique for Schmidt number $k \geq 4$ and therefore determining λ_k becomes computationally effortful. Nevertheless, choosing the operator X as for bipartite entanglement witnesses overcomes this problem, since there would be only one possible arrangement of the OS coefficients to a matrix yielding $\lambda_k = k - 1$. This shows that the fidelity-based Schmidt number witness is a special case of the OSD Schmidt witness with $X = |\Psi_+\rangle\langle\Psi_+|$. However, the fidelity-based Schmidt witness can not certify Schmidt number 3 for the example state ρ_3 but the OSD Schmidt witness can.

Bibliography

1. M, Weilenmann B. Dive D. Trillo E. A. Aguilar and M. Navascués. "Entanglement Detection beyond Measuring Fidelities". In: *Physical Review Letters* 124 (2020), p. 200502.
2. Ali Asadian. *private communication*.
3. C. Zhang S. Denker A. Asadian and O. Gühne. *Analyzing quantum entanglement with the Schmidt decomposition in operator space*. arXiv:2304.02447.
4. J. S. Frame G. de B. Robinson and R. M. Thrall. "The hook graphs of the symmetric group". In: *Canadian Journal of Mathematics* 6 (1954), p. 316.
5. J.-L. Basdevant. *Lectures on Quantum Mechanics*. Springer Nature, 2016. ISBN: 9783319434780.
6. H. J. Briegel and R. Raussendorf. "Persistent Entanglement in Arrays of Interacting Particles". In: *Physical Review Letters* 86 (2001), p. 910.
7. A. Sanpera D. Bruß and M. Lewenstein. "Schmidt number witnesses and bound entanglement". In: *Physical Review A* 63 (2001), 050301(R).
8. K. Chen and L. A. Wu. "A matrix realignment method for recognizing entanglement". In: *Quantum Information and Computation* 3.3 (2003), p. 193.
9. M. Lewenstein B. Kraus J. I. Cirac and P. Horodecki. "Optimization of entanglement witnesses". In: *Physical Review A* 62 (2000), p. 052310.
10. J.-A. Désidéri. "Multiple-gradient descent algorithm (MGDA) for multiobjective optimization". In: *Comptes Rendus Mathematique* 350 (2012), p. 313.
11. R. H. Dicke. "Coherence in Spontaneous Radiation Processes". In: *Physical Review* 93 (1954), p. 99.
12. A. K. Ekert. "Quantum Cryptography Based on Bell's Theorem". In: *Physical Review Letters* 67 (1991), p. 661.
13. S. J. Freedman and J. F. Clauser. "Experimental Test of Local Hidden-Variable Theories". In: *Physical Review Letters* 28 (1972), p. 938.
14. O. Gittsovich. "Quantum correlations: entanglement detection with second moments and the classical simulation of disordered systems". PhD thesis. Universität zu Innsbruck, 2010.
15. A. Aspect P. Grangier and G. Roger. "Experimental Realization of Einstein-Podolsky-Rosen-Bohm : A New Violation of Bell's Inequalities". In: *Physical Review Letters* 49 (1982), p. 91.
16. A. Aspect P. Grangier and G. Roger. "Experimental Test of Bell's Inequalities Using Time-Varying Analyzers". In: *Physical Review Letters* 49 (1982), p. 1804.

© The Editor(s) (if applicable) and The Author(s), under exclusive license to Springer Fachmedien Wiesbaden GmbH, part of Springer Nature 2023
S. Denker, *Characterizing Multiparticle Entanglement Using the Schmidt Decomposition of Operators*, BestMasters,
https://doi.org/10.1007/978-3-658-43203-4

17. A. Aspect P. Grangier and G. Roger. "Experimental Tests of Realistic Local Theories via Bell's Theorem". In: *Physical Review Letters* 47 (1981), p. 460.
18. O. Gühne and M. Seevinck. "Separability criteria for genuine multiparticle entanglement". In: *New Journal of Physics* 12 (2010), p. 053002.
19. O. Gühne and G. Tóth. "Entanglement detection". In: *Physics Reports* 474 (2009), p. 1.
20. Otfried Gühne. *private communication.*
21. M. Horodecki P. Horodecki and R. Horodecki. "Separability of mixed quantum states: linear contractions approach". In: *Open Systems & Information Dynamics* 13 (2006), p. 103.
22. M. Horodecki P. Horodecki and R. Horodecki. "Separability of mixed states: necessary and sufficient conditions". In: *Physics Letters A* 223 (1996), p. 1.
23. P. Horodecki. "Separability criterion and inseparable mixed states with positive partial transposition". In: *Physics Letters A* 232 (1997), p. 333.
24. R. Horodecki P. Horodecki M. Horodecki and K. Horodecki. "Quantum Entanglement". In: *Reviews of Modern Physics* 81 (2009), p. 865.
25. N. Johnston and D. W. Kribs. "Duality of Entanglement Norms". In: *Houston Journal of Mathematics* 41.3 (2015), pp. 831–847.
26. R. M. Adin R. C. King and Y. Roichman. "Enumeration of standard Young tableaux of certain truncated shapes". In: *The Electronic Journal of Combinatorics* 18 (2011), p. 20.
27. L. K. Shalm E. Meyer-Scott B. G. Christensen P. Bierhorst M. A. Wayne M. J. Stevens T. Gerrits S. Glancy D. R. Hamel M. S. Allman K. J. Coakley S. D. Dyer C. Hodge A. E. Lita V. B. Verma C. Lambrocco E. Tortorici A. L. Migdall Y. Zhang D. R. Kumor W. H. Farr F. Marsili M. D. Shaw J. A. Stern C. Abellán W. Amaya V. Pruneri T. Jennewein M. W. Mitchell P. G. Kwiat J. C. Bienfang R. P. Mirin E. Knill and S. W. Nam. "A strong loophole-free test of local realism". In: *Physical Review Letters* 115 (2015), p. 250402.
28. B. Kraus. "Entanglement properties of quantum states and quantum operations". PhD thesis. Universität zu Innsbruck, 2003.
29. O. Gühne M. Cuquet F. E. S. Steinhoff T. Moroder M. Rossi D. Bruß B. Kraus and C. Macchiavello. "Entanglement and nonclassical properties of hypergraph states". In: *Journal of Physics A: Mathematical and Theoretical* 47 (2014), p. 335303.
30. A. Acín D. Bruß M. Lewenstein and A. Sanpera. "Classification of Mixed Three-Qubit States". In: *Physical Review Letters* 87.4 (2001), p. 040401.
31. M. Bourennane M. Eibl C. Kurtsiefer S. Geartner H. Weinfurter O. Gühne P. Hyllus D. Bruß M. Lewenstein and A. Sanpera. "Experimental Detection of Multipartite Entanglement using Witness Operators". In: *Physical Review Letters* 92.8 (2004), p. 087902.
32. M. Curty M. Lewenstein and N. Lütkenhaus. "Entanglement as a Precondition for Secure Quantum Key Distribution". In: *Physical Review Letters* 92 (2004), p. 217903.
33. M. Curty O. Gühne M. Lewenstein and N. Lütkenhaus. "Detecting two-party quantum correlations in quantum-key-distribution protocols". In: *Physical Review A* 71 (2005), p. 022306.
34. R. Qu J. Wang Z.-s. Li and Y.-r. Bao. "Encoding hypergraphs into quantum states". In: *Physical Review A* 87 (2013), p. 022311.
35. V. Giovannetti S. Lloyd and L. Maccone. "Quantum Metrology". In: *Physical Review Letters* 96 (2006), p. 010401.
36. V. Giovannetti S. Lloyd and L. Maccone. "Quantum-Enhanced Measurements: Beating the Standard Quantum Limit". In: *Science* 306 (2004), p. 1330.

37. I. G. Macdonald. *Symmetric Functions and Hall Polynomials*. Oxford University Press, 1995, p. 267. ISBN: 0198534892.
38. O. Gühne Y. Mao and X.-D. Yu. "Geometry of faithful entanglement". In: *Physical Review Letters* 126 (2021), p. 140503.
39. B. Jungnitsch T. Moroder and O. Gühne. "Taming multiparticle entanglement". In: *Physical Review Letters* 106 (2011), p. 190502.
40. O. Nelles. *Nonlinear System Identification, From Classical Approaches to Neural Networks, Fuzzy Models, and Gaussian Processes*. Springer, 2021. ISBN: 9783030474386.
41. M. A. Nielsen. "Quantum Information Theory". PhD thesis. University of New Mexico, Albuquerque, 1998.
42. M. A. Nielsen and I. L. Chuang. *Quantum Computation and Quantum Information*. Cambridge University Press, 2010. ISBN: 9781107002173.
43. A. C. Doherty P. A. Parrilo and F. M. Spedalieri. "Distinguishing Separable and Entangled States". In: *Physical Review Letters* 88 (2002), p. 187904.
44. A. Peres. *Quantum Theory: Concepts and Methods*. Springer Dordrecht, 2002. ISBN: 9780306471209.
45. A. Peres. "Separability Criterion for Density Matrices". In: *Physical Review Letters* 77.8 (1996), p. 1413.
46. C. H. Bennett G. Brassard C. Crépeau R. Jozsa A. Peres and W. K. Wootters. "Teleporting an Unknown Quantum State via Dual Classical and Einstein-Podolsky-Rosen Channels". In: *Physical Review Letters* 70 (1993), p. 1895.
47. M. Piani and C. E. Mora. "Class of positive-partial-transpose bound entangled states associated with almost any set of pure entangled states". In: *Physical Review A* 75 (2007), p. 012305.
48. A. Pinciu. "On the number of shifted standard tableaux". In: *Bulletin mathématique de la Société des Sciences Mathématiques de la République Socialiste de Roumanie* 29 (1985), p. 285.
49. A. Einstein B. Podolsky and N. Rosen. "Can Quantum-Mechanical Description of Physical Reality Be Considered Complete?" In: *Physical Review* 47 (1935), p. 777.
50. O. Rudolph. "Further Results on the Cross Norm Criterion for Separability". In: *Quantum Information Processing* 4.3 (2005), p. 219.
51. O. Rudolph. "Some properties of the computable cross-norm criterion for separability". In: *Physical Review A* 67 (2003), p. 032312.
52. E. Schrödinger. "Die gegenwärtige Situation in der Quantenmechanik". In: *Naturwissenschaften* 23 (1935), p. 844.
53. J. Schur. "Über die Darstellung der symmetrischen und der alternierenden Gruppe durch gebrochene lineare Substitutionen". In: *Journal für die reine und angewandte Mathematik* 139 (1911), p. 155.
54. J. Shang and O. Gühne. "Convex Optimization over Classes of Multiparticle Entanglement". In: *Physical Review Letters* 120 (2018), p. 050506.
55. C. H. Bennett D. P. DiVincenzo T. Mor P. W. Shor J. A. Smolin and B. M. Terhal. "Unextendible Product Bases and Bound Entanglement". In: *Physical Review Letters* 82 (1999), p. 5385.
56. B. Hensen H. Bernien A. E. Dréau A. Reiserer N. Kalb M. S. Blok J. Ruitenberg R. F. L. Vermeulen R. N. Schouten C. Abellán W. Amaya V. Pruneri M. W. Mitchell M. Markham D. J. Twitchen D. Elkouss S. Wehner T. H. Taminiau and R. Hanson. "Loophole-free

Bell inequality violation using electron spins separated by 1.3 kilometres". In: *Nature* 526 (2015), p. 682.

57. B. M. Terhal and P. Horodecki. "Schmidt number for density matrices". In: *Physical Review A* 61 (2000), 040301(R).

58. R. M. Thrall. "A combinatorical problem". In: *Michigan Mathematical Journal* 1 (1952), p. 81.

59. W. Dür G. Vidal and J. I. Cirac. "Three qubits can be entangled in two inequivalent ways". In: *Physical Review A* 62 (2000), p. 062314.

60. H. Weinfurter and M. Żukowski. "Four-photon entanglement from down-conversion". In: *Physical Review A* 64 (2001), 010102(R).

61. M. Giustina M. A. M. Versteegh S. Wengerowsky J. Handsteiner A. Hochrainer K. Phelan F. Steinlechner J. Kofler J.-Å. Larsson C. Abellán W. Amaya V. Pruneri M. W. Mitchell J. Beyer T. Gerrits A. E. Lita L. K. Shalm S. W. Nam T. Scheidl R. Ursin B. Wittmann and A. Zeilinger. "Significant-Loophole-Free Test of Bell's Theorem with Entangled Photons". In: *Physical Review Letters* 115 (2015), p. 250401.

Printed in the United States
by Baker & Taylor Publisher Services